ISLANDS OF ORDER

Princeton Studies in Complexity

Series Editors
Simon A. Levin (Princeton University)
Steven H. Strogatz (Cornell University)

ISLANDS OF ORDER

A GUIDE TO COMPLEXITY MODELING FOR THE SOCIAL SCIENCES

J. Stephen Lansing and Murray P. Cox

Foreword by Michael R. Dove

PRINCETON UNIVERSITY PRESS

PRINCETON AND OXFORD

Published by Princeton University Press
41 William Street, Princeton, New Jersey 08540
6 Oxford Street, Woodstock, Oxfordshire OX20 1TR

press.princeton.edu

The epigraph to chapter 3 is quoted with the permission of the Estate of W. H. Auden.

Library of Congress Control Number: 2019942458
ISBN 978-0-691-19293-2
ISBN (pbk.) 978-0-691-19294-9
ISBN (e-book) 978-0-691-19753-1

British Library Cataloging-in-Publication Data is available

Editorial: Fred Appel, Thalia Leaf, and Jenny Tan
Production Editorial: Kathleen Cioffi
Text and Cover Design: Pamela L. Schnitter
Cover image courtesy of J. Stephen Lansing
Production: Erin Suydam and Brigid Ackerman
Publicity: Nathalie Levine and Kathryn Stevens

This book has been composed in Sabon LT Std and Ideal Sans

Printed on acid-free paper. ∞

Printed in the United States of America

10 9 8 7 6 5 4 3 2 1

For Herawati Sudoyo and Deborah Winslow

CONTENTS

FIGURES

TABLES

FOREWORD

The Malay Archipelago Revisited

> We necessarily read it not just as we please, but under circumstances not chosen by ourselves.
> —*Fernando Coronil, 1995, paraphrasing Marx on writing an introduction to a book.*[1]

Prehistory

For more than a century, anthropologists and others have studied, speculated, and debated about the peopling of mainland Southeast Asia, insular Southeast Asia, and the Pacific beyond.[2] Ongoing research uses studies of ancient DNA to examine farming expansion and language shifts, study of genetic turnover, and cultural transitions.[3] This research suggests that Pleistocene-era hunter-gatherers, Hoàbìnhians, populated insular Southeast Asia some 50,000 years ago. They were followed by Neolithic farming peoples from southern China—Kradai, Austroasiatic, and Austronesian speakers—who spread into mainland Southeast Asia some 5,000–4,500 years ago. The latter two groups reached the Malay archipelago approximately 4,100 years ago.

In the current work, J. Stephen Lansing and Murray P. Cox, an anthropologist and geneticist respectively, bring a unique perspective to this debate. They too are interested in asking what historical forces shaped the human societies of the Malay archipelago and the wider Pacific; but they abjure a focus on the dramatic sweep of large-scale population movement, culture change, and racial typologies. Much of the postulating regarding historic "migration" merely restates the reality, but does not explain it.

---

[1] F. Coronil. Introduction to *Cuban Counterpoint: Tobacco and Sugar*, by Fernando Ortiz. Duke University Press. 1995.

[2] I invoke the title of Alfred Russel Wallace's (1869) monumental nineteenth century work on the region, in recognition of not just the topics shared by the current volume but also its exuberant intellectual reach.

[3] P. Bellwood. 2018. The search for ancient DNA heads east. *Science* 361(6397): 31–32; M. Lipson, O. Cheronet, S. Mallick, et al. Ancient genomes document multiple waves of migration in Southeast Asian prehistory. *Science* 361(6397):92–95; H. McColl, F. Racimo, L. Vinner, et al. 2018. The prehistoric peopling of Southeast Asia. *Science* 361(6397):88–92.

Lansing and Cox seek to explain this history of movement in terms of the social dynamics of the small, persistent communities that have long characterized the region. They seek to explain large-scale transformations in terms of mundane, quotidian social practices like post-marital residence and inter-household coordination of labor. They are particularly interested in how languages have moved or not, and replaced others or been replaced themselves, across the archipelago.

Language

Some of the earliest theorizing about the historical connectedness of this region was prompted by the study of language, which is an old topic in the discipline of anthropology. The dominant figure in American anthropology in the early twentieth century, Franz Boas, carried out extensive research on the languages of the Inuit in the circumpolar north and the Native Americans of the northwest coast, and made the study of language into a central focus of the discipline. Antagonistic toward grand nineteenth-century theories on race, culture, and evolution, Boas (1940), in an early anti-essentialist stance, documented the many ways in which biology, language, and culture do not perfectly co-vary.[4]

The work by Lansing and Cox hearkens back to this tradition, but they bring new tools to bear upon it. They begin and end their volume with observations of the relatedness of languages across the Malay archipelago, beginning with Joseph Banks, who sailed with Cook in the eighteenth century; then Alfred Russel Wallace (1869) in the nineteenth century; and J.P.B. de Josselin de Jong early in the twentieth century at Leiden University. This relatedness has been much observed and discussed, but with some notable lacunae. It has been attributed to the migration of a homogeneous speech community across the region, but with little explanation of precisely how migrating peoples hang onto, or don't, common languages.

Lansing and Cox examine this question by looking at the relationship between language and genes at the community level, drawing on data sets from a sample of dozens of villages across the archipelago. Historical linguistics generally looks at changes in language over time, not in the speakers of language. Lansing and Cox flip this prioritization: they look at speech communities as opposed to speech. They look at the way that kinship practices—in particular, rules of descent and post-marital residence—play an active role in channeling language persistence and transmission. But they also show how easily languages can be given

[4] F. Boas. *Race, Language and Culture*. Collier-MacMillan Canada, 1940.

up, even when the language community in question persists. Indeed, they draw the profound and counterintuitive conclusion that communities of language speakers tend to outlast particular languages. This is reflected in the fact that some of the speech communities they study have shared genetic history that is far more ancient than any particular shared language.

Genetics

The application of new tools for studying DNA—paleogenomics—to the study of the peopling of the Indo-Pacific region is not without controversy. It has raised questions about race-based essentializing of identity. More generally, it has reenergized old debates over the determinant role in society of biology versus culture. This work by Lansing and Cox makes a powerful contribution to these debates. For example, their village-level study of population genetics demonstrates that Darwinian principles of male dominance rarely explain reproductive success. Indeed, they reinterpret the human significance of the "Wallace Line" based on spousal selection by women, not men.

The "Wallace Line" is the famous boundary that Wallace drew between the flora and fauna and people of the western and eastern parts of the archipelago, the "Indo-Malayan" and "Austro-Malayan divisions."[5] Whereas animals generally could not cross this line, migrating people with boats certainly could, but some didn't; there is a marked genetic differentiation from one side to the other. Lansing and Cox argue that the Wallace Line was a genetic barrier for humans for social, not geographic, reasons. One key factor was a slight tendency for Austronesian-speaking women to accept husbands from neighboring Papuan-speaking communities, which meant that their children would speak their mother's Austronesian language. Over a timescale of dozens of generations, this seemingly trivial deviation in marriage preferences produced seismic shifts in language, culture, and demography. This helps to explain the oft-noted mystery of why some Papuan groups adopted Austronesian languages. It exemplifies the approach taken by Lansing and Cox, in which social identity is a function of a complex interplay between genetic heritage, history, and social relations and rules.[6]

---

[5] A. R. Wallace. *The Malay Archipelago. The Land of the Orang-utan and the Bird of Paradise: A Narrative of Travel, with Studies of Man and Nature*. Periplus, 2000, pp. 7–15 (first published in 1869 by MacMillan).

[6] Cf. K. Tallbear. *Native American DNA: Tribal Belonging and the False Promise of Science*. University of Minnesota Press, 2013.

Emergent Complexity

The over-arching thesis of Lansing and Cox is that the complex patterns of linguistic and genetic diversity in the Malay archipelago are not random. They note that they began their study at a time when the long-reigning scientific paradigm of equilibrium studies was being turned upside down, so that scholars began to problematize not change but its absence. Aided by the development of theoretical tools like complexity theory and molecular genetics, Lansing and Cox set themselves the task of explaining not just change, therefore, but also stability. In a nonlinear world, they argue, stable equilibria—like persistent language communities—appear as islands of order in a sea of change, not a sea of order—hence, their title.

Lansing and Cox draw heavily on theories of nonlinear dynamics or complexity to explain how such order emerges. One of their case studies is the system of irrigated rice agriculture in Bali. Lansing (1991, 2006) has spent decades studying the rice terraces of Bali, in particular the fundamental tension between the need for water and the threat of pests.[7] The best way to share the scarce resource of water is to stagger the timing of the agricultural calendar, so that all farmers are not demanding water at the same time. But if everyone's cultivation time is staggered, so too will be the timing of their harvests, which will allow resident pest populations to concentrate on the ripening grain of each farm in turn. The best way to address the threat from pests, therefore, is to synchronize the timing of everyone's agricultural activities, so that the grain in all farms ripens at the same time, which then spreads resident pest populations thinly over an entire area. But synchronized timing means that everyone must demand irrigation water at the same time. Hence, there is a conundrum between the need to synchronize and the need to not synchronize.

Lansing and Cox, in this work, show how the Balinese strike an optimal balance between synchronization and non-synchronization, achieving an acceptable distribution of irrigation water and an acceptable burden from rice pests. This balance is achieved through the deliberation of cross-village irrigation organizations called *subak*, whose activities are ritually coordinated by a hierarchical complex of water temples. The role of the temples notwithstanding, this remarkable inter-household and inter-village system is not orchestrated top-down, it is locally generated. It

---

[7] J. S. Lansing. *Priests and Programmers: Technologies of Power in the Engineered Landscape of Bali.* Princeton University Press, 1991; J. S. Lansing. *Perfect Order: Recognizing Complexity in Bali.* Princeton University Press, 2012.

is an example of what Lansing termed in an earlier work "emergent complexity," here called "adaptive self-organized criticality" or, felicitously, "order for free." Within the Malay archipelago, this system is unique to Bali; it is absent on Java, despite Java's own deep history of irrigated rice cultivation. Lansing and Cox suggest that the great hydraulic kingdoms of medieval Java provided the high-level control to its system of rice agriculture; whereas the development of bottom-up control of agriculture in Bali helps to explain why its own indigenous kingdoms did not develop to the same extent.

This phenomenon of emergent complexity is an apparent logic without an overarching logician, a design without a designer. The overall logic of the irrigation system in Bali is nowhere articulated except symbolically in the system of ritual water temples. In their study of it, Lansing and Cox thus address one of the thorniest problems in the study of social order: how does it arise, and how do we talk about the process by which it arises. The lack of a welcoming niche for such order in the western development mind-set led to the imposition of the green-revolution model of rice cultivation in Bali in the 1970s, forcing the cessation of the coordination of agricultural schedules by the *subak* and water temples, which quickly led to an explosion of pest populations and a collapse in harvests—a mistake that was eventually recognized by the agricultural development experts.

An earlier generation of environmental anthropologists tried to explain such systems of environmental management in terms of an external, etic, or "operational" model, which contrasts with the internal or emic or "cognized" one.[8] Lansing and Cox greatly improve upon this explanation—unsatisfactory in either an external or internal sense—with their concept of self-organized criticality, which approaches in many respects the native Balinese view of the way that the system operates, at the same time as it can cross the cultural divide to be understood by outsiders.

Summary

Islands of Order is an extraordinary book, based on more than a decade of interdisciplinary research, involving dozens of villages and multiple islands across the Malay archipelago. It is interdisciplinary and inductive, driven not by a preformulated agenda but by intense intellectual curiosity regarding the region's history. Elegantly spanning a number of different

---

[8] R. A. Rappaport. *Pigs for the Ancestors: Ritual in the Ecology of a New Guinea People*. Yale University Press, 1968, pp. 237–42.

disciplines ranging from paleogenomics to ethnography, it stands alone in linking interest in questions regarding the "deep" past and the "thick" present. *Islands of Order* will be of keen interest not only to southeast Asianists, but to all students of the history of human societies.

Michael R. Dove
Yale University
3 February 2019

PREFACE

"In the 1950's and 1960's," Adam Kuper tells us, "the social sciences were better funded, better organized and generally in better spirits than ever before."[1] Much of this enthusiasm can be attributed to the influence of Talcott Parsons, who proposed a unified approach to the social sciences in which each of the disciplines—psychology, sociology, economics, even anthropology—would play its part. The application of consistent statistical methods would guarantee steady progress toward a grand synthetic theory. But the popularity of Parson's vision began to fade by the 1970s, leaving, in the words of biologist David Sloan Wilson, "a vast archipelago of disciplines that only partially communicate with each other."[2]

At about the same time, faltering steps were taken in the direction of a different kind of unification, inspired not by dreams of a grand theory, but rather by dissatisfaction with existing models. Computers made it possible to ask new questions about systems with inherent instability, like weather, epidemics, or the stock market. At the Santa Fe Institute and elsewhere, researchers found that simulation models developed for one domain could often be extended to others. Not infrequently, it turned out that broadening the scope of models increased not only their range, but also their explanatory power. One could, for example, explore the dynamics of earthquakes, stock crashes, and extinction events by modeling avalanches in sand piles.

What caught people's attention in these simulations was the frequent appearance of dramatic, nonlinear transitions—sharp jumps between otherwise stable states. In retrospect, perhaps we should not have been surprised: the difference between linear and nonlinear systems is precisely the possibility for new, unexpected patterns to emerge. A linear relationship is one in which effects are proportional to their causes: more of A produces more of B. This makes for easy calculations and simple, tidy models. But as Enrico Fermi quipped, "it does not say in the Bible that all laws of nature are expressed linearly."[3] While all linear systems are solvable analytically, most nonlinear systems are not, and until the advent of computers, little was known about them.

---

[1] A. Kuper. *Culture*. Harvard University Press, 2000, p. 15.
[2] D. S. Wilson. *Darwin's Cathedral*. University of Chicago Press, 2002, p. 83.
[3] J. Gleick. *Chaos*. Penguin, 1988, p. 68.

Still, interest in nonlinear systems was slow to catch on among social scientists. Initially there was skepticism about the relevance of exotic nonlinear processes to the everyday world of human experience. Perhaps the weather is subject to butterfly effects, but surely societies are more predictable? "[P]ositions of unstable equilibrium," wrote the famous economist Paul Samuelson in 1983, "even if they exist, are transient, non-persistent states.... How many times has the reader seen an egg standing on its end?"[4]

And yet, while economies and societies may appear to be stable, they do occasionally undergo fundamental, nonlinear change. If we try to analyze such changes from the perspective of standard equilibrium theory, what will we see? The answer is not the emergence of novel conditions. Because linear models explain steady progressive change, nonlinear data viewed from a linear perspective usually appear as nothing more than noise. Interestingly, most social science data are noisy. It also exhibits extraordinary diversity, for which there is no easy explanation from the Parsonian perspective. In a nonlinear world, stable equilibria would appear as islands of order in a sea of change.

Which brings us to this book. It is easy to build a case that we should keep an eye out for nonlinear interactions. The problem is how to go about it. What brings relatively stable patterns—islands of order—into existence? And what causes them to change? We begin with an introduction to models of change, from simple random drift to coupled interactions, phase transitions, co-phylogenies, and adaptive landscapes. Subsequent chapters adapt these general models to address specific questions in a series of case studies. These are drawn from the islands of the Indo-Pacific region, where the authors—an anthropologist and a geneticist—have worked for many years. They range across a variety of topics, like dominance, co-evolution, and cooperation, and across diverse scales of space and time. Although the case studies can be read separately and in any order, the questions progress from larger to smaller scales, and a key message is the importance of scale for bringing change into focus.

We set the stage with our first example, which concerns a cultural pattern that occurs at the largest geographic scale, and triggered the first anthropological question to be asked about the region. It was posed by Joseph Banks, the first head of the Royal Society, when as a young man he sailed on Cook's first voyage into the Pacific. As they traveled through the islands, both Banks and Cook were surprised to find that the languages of Polynesia were closely related. In his journal, Banks listed Tahitian and New Zealand Māori words side by side to show that the

---

[4] P. A. Samuelson. *Foundations of Economic Analysis.* Harvard University Press, 1983 (original 1947).

two languages are nearly identical. Banks used the same method for comparing vocabulary to trace a linguistic relationship westward all the way to Southeast Asia.[5] Later, Banks learned that the relationship extends much further, across the Indian Ocean to Madagascar. In September 1771, Banks wrote in his journal "how any Communication can ever have been carried between Madagascar and Java ... is I confess far beyond my comprehension."

Yet Banks was correct. His word lists identified a language family that is now called Austronesian. Subsequent discoveries showed that the Austronesian languages were spread by colonists from mainland Asia, whose voyages took them through the islands of the Malay archipelago. The Austronesians brought with them their languages, religions, kinship systems, material culture, domesticated animals, diseases, and DNA. As their colonies took root in the islands, all of these things began to change. The question of how and why such changes occurred invokes the second meaning of our title. Over the past two decades, the authors have studied dozens of Austronesian-speaking communities on seventeen islands from Borneo to New Britain. We used linguistic markers and neutral genetic variants to shed light on the demographic histories of these communities. These markers gave us a generational timeline, a baseline for tracking changes in language, culture, and social life. Some changes were dramatic, like a switch from matrilineal to patrilineal systems of descent. Others, equally profound, were more gradual, like the appearance of new languages. Cumulatively, these studies suggest that change is seldom either continuous or chaotic. Instead, it often takes the form of a transition from one relatively stable state to another—a journey between islands of order.

In retrospect, we were fortunate both in our timing and in the sites we chose to investigate. A generation ago, anthropologists studying nonliterate societies could learn little about the past. Across the social sciences, explanatory models focused on equilibrium states. That began to change with the flowering of two new fields: molecular genetics, and nonlinear dynamics or "complexity." Molecular clocks made it possible to trace the ancestries of men and women, and new analytical methods allowed

---

[5] "The Madagascar language has also som[e] words similar to Malay words, as ouron the nose, in Malay Erung Lala, the tongue Lida Tang, the hand Tangan Taan, the ground Tanna. From this similitude of language Between the inhabitants of the Eastern Indies and the Islands in the South Sea I should have venturd to conjecture much did not Madagascar interfere; and how any Communication can ever have been carried between Madagascar and Java to make the Brown long haird people of the latter speak a language similar to that the Black woolly headed natives of the other is I confess far beyond my comprehension." J. Banks. *The Endeavour Journal of Sir Joseph Banks* 1768–1771. 1st edition. Angus and Robertson, 1962.

out-of-equilibrium dynamics to be recognized in physical and biological systems. In anthropology, molecular clocks were first used to study events in deep time, such as the divergence of hominins from apes. But the resolution of the clocks improved quickly, and it has recently become possible to track change on a time scale of a few generations, which happily encompasses the relevant period for most social science.

We were also lucky to be working in the islands, where the Austronesian colonists tended to settle down, stay put, and adapt to local conditions. Most lived in small, persistent communities, where innovations of all sorts, from language to kinship, material culture and disease resistance, were seldom erased by the passage of time. These were ideal conditions for our generational timelines, because most changes occurred in situ. When that happens, the result can be pictured as a tree: a branching phylogeny with roots in the past and tips in the present. This framework applies to languages as well as to genetic ancestry, and often to other heritable traits like material culture. But trees are not inevitable: over time, mixing caused by a mysterious process (sex) gradually scrambles the traces of ancestry in most of our DNA. Languages and artifacts, too, are passed around or vanish as a result of population movements.

But on these islands, neutral genetic markers—strings of DNA that do not change our looks or behavior, but retain a record of our past—indicate that entire communities tended to stay intact over many generations after their initial colonization. That makes it possible to ask questions about processes that occur within communities: the scale at which nearly all social life occurs, where people learn languages, marry, bear children, compete for social rank, suffer and recover from disease, migrate or stay at home. A surprising amount of this behavior has to do with sex, and thus with genetics, and all of it leaves traces that can be pegged to generational timelines. To recover this information, we shifted our analytical focus from individuals (the usual target of genetic research) to groups of neighboring communities, from one island to the next.

Take one example: Over time, kinship systems and marriage customs leave a signature in the genetic composition of communities. If women remain in their home village and attract husbands from elsewhere, the results are readily apparent in the genetic diversity of the community. Conversely, if the men stay put and women marry in, a different pattern is created. If high-ranking men have more wives and children, yet other signals appear in the genetic composition of the community. Change always leaves a signature: chance events generate a characteristic pattern that is easily distinguished from changes caused by evolution or purposeful behavior. Overall, as we came to see, communities are shaped by processes that function at different scales, run at different rates, and, most importantly, interact.

On one level, then, this book consists of a collection of anthropological stories about South Sea islanders (for this we make no apology, it's our job and someone has to do it). But the book is organized to also facilitate a different sort of reading: the study of change itself. Along that path, our argument rests on two premises. The first is that cultures, languages, and societies, like other living systems, are seldom at equilibrium.[6] The second is a consequence: nonlinear interactions should be expected. As we will see, an impressive array of new analytical tools have recently become available to bring such interactions into focus. Typically, these methods were initially developed to address questions in fields like genetics or physics, and they may at first seem to have little relevance beyond those specific realms. But as our case studies show, at a deeper level they offer complementary insights into patterns that would otherwise remain hidden. A generation ago, complexity researchers began to investigate the quotidian patterns triggered by butterfly wings, dripping faucets, tumbling sand piles, and simple games. Broadening the analytical frame to encompass nonlinear dynamics opened new vistas in physics and biology. A similar opportunity beckons in the social sciences.

Plan of the book

As we mentioned above, the "islands of order" in our title are meant to refer both to attractors in the mathematical sense and to places where sea breezes rustle the palms. One of these narrative threads is progressive: in the chapters to follow, we construct a series of models that begin with the simplest models of change, and progress to greater depth and complexity. The second theme—the anthropology of the islands—is a little more complicated. In the conclusion to the book, we assemble the models into a comprehensive analysis of the historical forces that shaped the societies of the Malay Archipelago. But our models are not exclusively tethered to the anthropology of this region; they aim to reveal the warp and weft of the fabric of societies, and to a varying degree should hold for many places, peoples, and times. Here is how we will weave these themes together.

Chapter 1 Models of Change
   This chapter provides an introduction to the concepts and models of change we will use in the book. The key models are also implemented on our website, so that readers can try them out, at https://www.islandsoforder.com.

[6] This argument is not original with us; see I. Prigogine. *Self-Organization in Non-Equilibrium Systems*. Wiley, 1977.

Models: A general introduction and overview, with functioning models of logistic growth and Daisyworld available on the book's website.

Chapter 2 Discovering Austronesia

Here we set the geographic and historical stage with a history of the colonization of Island Southeast Asia. We pick up the story with the question posed by Joseph Banks: how is it possible that the languages of isolated islands located more than half a world apart could be closely related? In the nineteenth century, two naturalists, Eugène Dubois and Alfred Russel Wallace, proposed models of successive waves of migrations. Subsequently, archaeologists and historical linguists clarified this picture. But the advent of genetic research in the late twentieth century challenged Wallace's model of steady progressive colonization. In 2010, we discovered that one of the greatest geographic barriers to the flow of genes between human populations, equivalent to the Sahara or the Himalayas, occurs in the midst of a continuous chain of islands that forms the southern arc of the Malay archipelago. This discovery prompted new questions about the varying roles of women and men in the colonization process. Combining genetics with linguistics and ethnography, in this chapter we develop a simple model of sex-biased migrations that offers an intriguing answer to Banks's question.

Model: We use a simple linear model to investigate the cumulative demographic, genetic, and cultural consequences of intermarriage between waves of colonists.

Chapter 3 Dominance, Selection, and Neutrality

We dig deeper into the genetic traces left by the colonization process, in particular the social processes that shape the demographic composition of villages. Did the sex bias that began in the early waves of colonization leave enduring traces in the demographic composition of villages? Is there evidence that men successfully competed for social dominance? To find out, we look for traces of past competition for social dominance and reproductive success. We use analytical methods and perspectives from two schools of evolutionary theory: population genetics and human behavioral ecology. However, these two approaches do not match seamlessly. The key question is how to distinguish change caused by selection from change caused by random drift. We step back to analyze the underlying theories of change, and show how the results and methods are relevant not only to our islands, but to ecology, genetics, archaeology, and sociology.

Models: Neutrality and selection for genetics, species, artifacts, and children's names.

Chapter 4 Language and Kinship in Deep Time

Chapter 3 showed that cultural rules governing marriage leave clear signatures in the demographic composition of communities. How long do these signatures persist, and what are the broader consequences for social life? In this chapter we shift our perspective from questions of dominance and selection to the role of kinship systems in channeling language, culture, and social relationships between groups. In *On the Origin of Species*, Charles Darwin proposed that human races and languages evolved in concert following a tree-like history of splits and isolation. But later studies showed that the association between genes and languages is often transient. In search of deeper patterns, we investigate what happens when genetic and linguistic trees become joined by kinship systems at the community scale in a study of 25 villages on two islands speaking 17 different languages. This is the third of three chapters about the tribal societies located east of the Wallace Line.

Models: Language trees, sex-biased migration, and cophylogenies of languages and genes, channeled by kinship.

Chapter 5 Islands of Cooperation

In this chapter we turn our gaze west of the Wallace Line to the rice-growing villages of Bali, seemingly a world away from the tribal societies to the east. The gods favored Bali with nearly ideal conditions for growing paddy rice, and the Balinese began to construct irrigation tunnels and rice terraces on the slopes of their volcanoes late in the first millennium AD. Farmers became deeply invested in their engineered landscape, and to preserve this inheritance, marriage to the girl next door became the norm. This brought population movement across the Wallace Line largely to a standstill.

In this, the first of three chapters about social-environmental interactions, we investigate the dynamics of cooperation on Bali. We focus on the emergence of a fragile system of cooperative management by farmers, which sustained an equally fragile infrastructure of terraces, aqueducts, and irrigation tunnels for many centuries. What caused these islands of cooperation to emerge, what sustains them, and why do they sometimes fail? These questions are taken up successively in chapters 5, 6, and 7. The stage is set in chapter 5 with an introduction to the origins, ethnography, and ecology of *subaks*, the social institutions that, over the course of about a thousand years, terraced the slopes of Bali's volcanoes.

*Model: Coupled social-ecological systems: a two-player coopera-
tion game and a systems dynamical model of 173 subaks along two
Balinese rivers.*

Chapter 6 Adaptive Self-Organized Criticality

How do higher-level systems of control emerge from the local
interactions of subaks? This chapter focuses on a hitherto-
unknown phase transition that occurs as cooperation spreads,
which leaves traces visible in satellite imagery. Because similar
traces should exist in other anthropogenic landscapes, this analysis offers new insights into the evolution of social-environmental
couplings.

Model: Phase transitions and adaptive self-organized criticality.

Chapter 7 Transition Paths

Why are some subaks more successful than others? And what
influences the transition between more and less successful regimes?
To find out, we surveyed farmers in two dozen villages with simple
questionnaires, and adapted methods from information geometry to
interpret the results. These methods offer a way to search for multiple attractors, investigate how they differ, and calculate the most
likely transition paths between them. Once again, these methods are
not tied to Bali.

*Model: Transition paths between regimes on an adaptive land-
scape.*

Chapter 8 From the Other Shore

We conclude by engaging the two themes of the book in a dia-
logue about change. To do so, we revisit the Austronesian expansion
and theoretical ideas developed by anthropologists to explain why
some features of language, culture, and social organization resisted
change. Next we revisit the models used in earlier chapters to analyze trajectories of change. Each chapter uses one or more models to
investigate a specific question, yielding snapshots of features of the
islands. Like all snapshots, these do not yet fit neatly together into a
single montage, though one is quickly emerging. We show that they
can be combined in ways that engage decades of debates about the
anthropology of the islands. We conclude by going beyond those
debates to explore the consequences of the phase transition that set
Balinese society on a new course.

*Models: Phase portraits of earlier models, combined to explore
the interaction of kinship with social organization, language, dominance, and genetics.*

Models: Our online resource site for *Islands of Order*

The models described in each chapter can be explored and run using our online resources at https://www.islandsoforder.com. While this book contains the theory, we hope you will get your hands dirty and explore how the models really work. Our online apps give you everything you need to do so.

Acknowledgments

Each chapter draws on our previously published research. We have tried to distill the key ideas and show how they are connected. Most publications were co-authored, and our immediate debts to our co-authors and sponsoring institutions are acknowledged in those publications. But the intellectual debt we owe to our co-authors extends much further: findings from each phase of fieldwork and modeling became the starting point for new investigations. However, the views expressed in this book are our own and may not reflect those of our colleagues and co-authors. In particular, we give our thanks to Cheryl Abundo, Wayan Alit Arthawiguna, Tanmoy Bhattacharya, Ginger Booth, Lock Yue Chew, Ning Ning Chung, Yves Descatoire, Michael Dove, Sean Downey, Elsa Guillot, Brian Hallmark, Michael Hammer, Georgi Hudjashov, Guy Jacobs, Tatiana Karafet, James Kremer, Safarina Malik, John Miller, John Murphy, Peter Norquest, Yancey Orr, Vernon Scarborough, Ashley Stinnett, Herawati Sudoyo, Hendrik Sugiarto, Sang Putu Kaler Surata, Stefan Thurner, Meryanne Tumonggor, Helena Suryadi, Joseph Watkins, and Deborah Winslow. We personally acknowledge our late colleague, Olga Savina.

We also express our profound gratitude to the following institutions whose support made this book possible: Nanyang Technological University; Massey University; the Santa Fe Institute; the Eijkman Institute of the Indonesian Ministry of Research, Technology and Higher Education; the Max Planck Institute for Evolutionary Anthropology; the Singapore Ministry of Education; the Alexander von Humboldt Foundation; and the US National Science Foundation. Finally, we gratefully acknowledge the inspiration of two great scholars of Austronesia, James J. Fox and Peter S. Bellwood.

ISLANDS OF ORDER

~~~~~~

CHAPTER 1

Models of Change

> Order is Heav'n's first law.
>
> —*Alexander Pope*

> Quid sit prius actum respicere aetas
> nostra nequit nisi qua ratio vestigia monstrat.[1]
>
> —*Lucretius*

A point of departure

In this chapter, we provide the theoretical background for the chapters to follow. We begin with an overview of models of change—the same models that we apply in later chapters to specific cases. Here our goal is to equip the reader with a feel for the possibilities. Those possibilities look very different now than they did a generation ago. In the nineteenth century, change was the great topic of social theory, but by the mid-twentieth century, it had largely ceased to have analytical importance for the social sciences. Our point of departure is a brief exploration of the reasons why this occurred. Interestingly, those reasons vary from one field to the next. What role did change play in social and evolutionary theory circa 1965, before the discovery of deterministic chaos and the molecular revolution in genetics?

Anthropology and sociology

In 1962 Claude Lévi-Strauss published a profound challenge to the received wisdom of anthropology concerning the significance of cultural change. Nineteenth-century anthropology inherited a view of change based on stadial models, with roots that extend into classical antiquity. Between 700 and 300 BC, the ancient Greeks developed a theory of the evolution of human societies that persisted for more than two millennia. In these stadial models, societies are propelled from one stage to the next by innovations such as fire, cereal cultivation, language, metallurgy, and writing. Roman writers like Lucretius continued this speculative tradition, and in the seventeenth and eighteenth centuries, philosophers like

---

[1] "What came before, our age cannot look back to, except insofar as reason shows the traces." Titus Lucretius Carus, *De Rerum Natura*, 5.1446–47.

Condorcet, Hegel, Comte, and others used stadial models to champion the progress of rational thought. Anthropologists such as Tylor, Morgan, and Spencer later sought to embed these models within an evolutionary framework. From this perspective, change is intrinsically purposeful; as Marx put it, "the five senses are the work of all previous history... history is the true natural history of mankind."[2]

Lévi-Strauss's challenge to this view was a logical extension of his structuralist program in language to encompass other cultural phenomena. In the influential structural theory of language developed by Ferdinand de Saussure, knowledge of the prior state of a language tells us nothing about its present workings. Lévi-Strauss adapted Saussure's methods to create a powerful and predictive theory of how culture is organized by symbolic systems, affecting everything from cooking to cosmology. In culture as in language, argued Lévi-Strauss, knowledge of the antecedents of symbols is irrelevant to their current meaning. And more forcefully, the effect of change is to shatter the internal consistency of these systems of thought. Traditional societies seek "to make the states of their development which they consider 'prior' as permanent as possible. ... There is indeed a before and an after, but their sole significance lies in reflecting each other."[3]

Lévi-Strauss's structuralist challenge subsequently loomed large in anthropology. As its influence grew, change came to be seen as little more than a source of disorder. Efforts to improve stadial models of culture and society were marginalized.

Economics

Economics took a similar path for different reasons. In the 1950s, economists proposed that the self-regulating capacity of market-based economies is founded on a state of general equilibrium. By the 1960s, the key theoretical questions in economics were the stability of equilibria to shocks and how the economy transitioned between equilibria. In considering both questions, economics takes the perspective that economies are either at equilibrium, returning to equilibrium following a perturbation, or heading toward a new equilibrium.

Genetics

As in economics, much early evolutionary theory also focused on the attainment of equilibrium. But the biologist's concept of equilibrium was different from that of economists. In 1930, Ronald Fisher introduced

---

[2] K. Marx. *Economic and Philosophic Manuscripts of 1844*. Penguin, 1974, p. 136.

[3] C. Lévi-Strauss. *The Savage Mind*. University of Chicago Press, 1966, p. 234.

Boltzmann's model of statistical equilibrium into genetics. According to Fisher's *Fundamental Theorem of Natural Selection*, "[t]he rate of increase in fitness of any organism at any time is equal to its genetic variance in fitness at that time."[4] In this view, natural selection operates on populations of organisms with varying fitness, propelling them toward fitness peaks that are statistical equilibria. These different definitions of "equilibrium" had analytical consequences: for general equilibrium in economics, any points lying away from equilibrium are merely transient states, and thus safely ignored, while in biology, Fisher's theorem proposes that variation provides the raw material for selection, determining the rate of change.[5]

Thus by the 1960s, the old stadial conception of change propelled by innovation was nearly forgotten across the social and evolutionary sciences, except by archaeologists and Marxist historians. For the equilibrium models that took its place, change was merely a transient state, and in some readings, a source of disorder. The new equilibrium models came in two forms. In economics, classical or Newtonian equilibrium meant the solution to a system of coupled differential equations.[6] In genetics, Boltzmannian statistical equilibrium described the average state of a collection of particles. Both concepts of equilibrium—classical and statistical—required external forces to shift from one equilibrium state to another.[7]

But this assumption only holds for linear systems. Nonlinear systems differ from Boltzmannian statistical ensembles in that initial differences may not average out. Instead, outliers can initiate large-scale spontaneous reorderings and movement to new attractors. As research on nonlinear dynamics continued, it became clear that spontaneous self-organization can achieve many of the outcomes traditionally assigned to impinging forces.[8] The mathematics showed, intriguingly, that self-organization can cause qualitative change in the behavior of dynamical systems, as Ilya

---

[4] R. A. Fisher. *The Genetical Theory of Natural Selection.* Clarendon Press, 1930, p. 35.

[5] In this era, geneticists debated the significance of variation: is it a "genetic load"—that is, a burden imposed by misreadings in our genes—or is it instead the raw material from which we benefit evolutionarily as our environments change?

[6] Later, some economists began to reframe their models as statistical equilibrium, but "...the concept of statistical equilibrium remained unknown to most economists throughout all the XXth century and up to now." U. Garibaldi and E. Scalas, 2010 Tolstoy's dream and the question for statistical equilibrium in economics and the social sciences. In G. Naldi, L. Pareschi, and G. Toscani, eds. *Mathematical Modeling of Collective Behavior in Socio-Economic and Life Sciences.* Springer, p. 116.

[7] In genetics, these forces were caused by natural selection; in economics, by changes in the parameters affecting price.

[8] S. Kauffman. *The Origins of Order: Self-Organization and Selection in Evolution.* Oxford University Press, 1993.

Prigogine observed in his 1977 Nobel address. By the 1980s, the origins of order in nonlinear dynamical systems had moved to the forefront of research on complex systems. Biologist Robert May commented that "even in vastly complicated interactive networks, a few simple rules can easily—if amazingly—lead to order and self-organised patterns and processes. This represents a major advance in understanding how the living world works."[9]

The origins of order

May's reference to interactive networks reflects a shift in perspective on population structure. In economics, the concept of general equilibrium describes populations of economic actors engaged in buying and selling, whose actions depend on the state of the market. In genetics, following Fisher's model, chance mutations generate variation in populations of organisms, which in turn provides the raw material for natural selection. In both cases, a population is in effect a cloud of points. Clouds can drift, change shape, and become more or less dense, but they have no internal organization. In contrast, populations as networks have different properties than populations as clouds; points (or nodes) connected into networks can interact in vastly more complex ways. As another leading biologist, Richard Lewontin, commented, "[t]he facile claim that natural selection can accomplish every adaptive change fails to grapple with the problems posed by a highly structured system with its own laws of assembly and interaction."[10]

The image of populations as "vastly complicated interactive networks"[11] soon became an empirical reality for geneticists, as new technologies made it possible to decode genes and regulatory systems. In 1965 Jacques Monod received the Nobel Prize for describing the first example of a gene regulatory network. Five years later, in a book-length essay, *Chance and Necessity*, Monod argued that "chance alone is at the source of all novelty, all creation in the biosphere."[12] But progress in his own field soon contradicted this view. When it became possible to directly observe many gene regulatory systems, chance rapidly gave way to necessity. In time, the methods developed by geneticists to assess the role of chance began to be applied in other fields, from ecology to

[9] Robert M. May, in a 1993 review of Kauffman's book published in *The Observer*.

[10] Richard C. Lewontin, in a back-cover endorsement of Kauffman's 1993 book.

[11] May, 1993 book review.

[12] J. Monod. *Chance and Necessity*. Vintage, 1972, p. 112 (first published in 1970 as *Le Hasard et la Nécessité*).

linguistics to archaeology. The combination of better empirical data and a greater theoretical understanding of nonlinear dynamics revived interest in change, and shifted the analytical focus from anonymous individuals to evolving networks, populations, and communities.

This intellectual pirouette merits careful attention. We will trace it by analyzing a series of key discoveries that brought unanticipated patterns of emergent behavior into focus. In the second part of this chapter, these threads are drawn together into a comprehensive framework that provides the starting point for the case studies. We begin with the discovery of molecular clocks, which made it possible to directly measure rates of change in evolving systems.

A clock that keeps good time

DNA is composed of strings of nucleotides—As, Cs, Ts, and Gs. Only part of this DNA represents genes and thus contains the blueprints for making proteins, but all nucleotides, even the ones that have no physical or behavioral effects, are subject to mutation. These mutations may change just one letter, or entire strings of letters. For many parts of our DNA, especially outside the genes, mutations accumulate at a steady rate, with that rate conceptually mimicking the ticking of a clock.

The possibility that such clocks might exist was anticipated before it became possible to observe them directly. In 1967, biologist Allan Wilson and his student Vincent Sarich published a paper in *Science*, in which they suggested that the origin of the human species could be dated by means of the genetic mutations that have accumulated since humans and chimpanzees last shared a common ancestor. At the time, it was not possible to actually count those mutations. Instead, Sarich and Wilson compared diversity in serum albumen (a blood protein) among a large number of primates, and found that "[l]ineages of equal time depth show very similar degrees of change in their albumins. The degrees of change shown would therefore seem to be a function of time." Using this "evolutionary clock or dating device," they estimated that humans and chimpanzees diverged around five million years ago.[13]

This proposal was met with profound skepticism by most anthropologists, who favored a date of about 25 million years based on the fossil record of the time. Donald Johanson's discovery of the fossil hominin Lucy in 1974 provided compelling support for this younger chronology, but arguments about the validity of the "molecular clock" concept

[13] V. M. Sarich and A. C. Wilson. 1967. Immunological time scale for hominid evolution. *Science* 158:1200–3.

continued until the 1980s, when it became possible to sequence DNA and read off the mutations directly. That made it possible to find as many molecular clocks as one liked—from regions of the DNA that tick slowly to regions that tick fast. This discovery conclusively validated the method of molecular dating, and confirmed Sarich and Wilson's estimate of a five-million-year-old origin of our species. Today molecular clocks are no longer controversial (albeit more sophisticated); they are everyday tools in population genetics, and play a role in about half of the case studies in this book. But the very accuracy of molecular clocks triggered a new controversy: if mutations are regular, does this weaken the role of selection in the evolution of DNA?

Neutral drift

A year after the publication of Sarich and Wilson's paper on molecular clocks, geneticist Motoo Kimura predicted that the vast majority of evolutionary changes at the molecular level are caused not by selection, but by chance: the random drift of selectively neutral mutants. Even in the absence of selection, Kimura reasoned, evolutionary change will occur as a result of chance, and this could be analyzed with tools from probability theory. The idea that selection might have little or no role in shaping portions of the genome was not altogether new: in a famous disagreement with Ronald Fisher, Sewall Wright emphasized the importance of neutral processes such as drift as early as the 1930s. But Kimura took this idea further, offering a probabilistic method that can readily test for selective effects using data from the genome.

In genetics, the neutral theory was hotly debated for decades. As Kimura observed in his 1968 paper, the prevalent view in the 1960s held that almost all mutations are under selection, and this opinion was slow to change. But as Stephen J. Gould wrote in 1989, "[t]hese equations give us for the first time a baseline criterion for assessing any kind of genetic change. If neutralism holds, then actual outcomes will fit the equations. If selection predominates, then results will depart from [neutral] predictions."[14] This eventually led to a dramatic reversal in the way selection is viewed in molecular biology: geneticists now infer selection only when it can be shown that the assumption of neutrality has been violated. The success of the neutral theory triggered a shift in perspective, from the fitness of individual units of selection to the population-level consequences of both selection and drift.

[14] S. J. Gould. 1989. Through a lens, darkly: Do species change by random molecular shifts or natural selection? *Natural History* 98:16–24.

But is the neutral theory relevant above the molecular level? Theoretical ecologists began to consider this question in the 1990s. Previously, the prevalence of species in ecological communities was approached from a pan-selectionist perspective too: what are the special attributes of each species that explain its abundance in a given environment? Neutral theory offered an alternative hypothesis. If one assumes that species do not differ in their competitive abilities, what would the prevalence of species be if this depended only on the size of the total ecological community and the chance arrival of new species? In other words, do neutral processes of drift and replacement largely govern the formation and persistence of ecological communities? This question became one of the most hotly debated topics in theoretical ecology.[15] Mathematically, the neutral theory in ecology is faithful to its origins in genetics; both rely on the same underlying mathematical model.

Although the scope of the neutral theory in ecology is still being tested, a shift is underway from the assumption of pan-selectionism to the view that selection can only be inferred by showing departure from a null model of neutrality.[16] As in genetics, this represents a change in the level of analysis, from the fitness of individuals to the effects of selection at the community level. As Kimura wrote in 1983, "it is easy to invent a selectionist explanation for almost any specific observation; proving it is another story. Such facile explanatory excesses can be avoided by being more quantitative."[17]

Nonlinear systems

Kimura's linear equations for neutral drift have marvelous predictive power because there is only one neutral frequency distribution for any given population, depending solely on the mutation rate and the population size. (Thus, if we view genetic types or species of tree as a bag of marbles, the equilibrium distribution of colors reflects only the number of marbles in the bag and the rate at which new colors appear.) This is also true for the adaptations of Kimura's model in ecology and

[15] J. Harte. 2003. Tail of death and resurrection. *Nature* 424:1006–7; D. Alonso, R. Etienne, and A. McKane. 2006. The merits of neutral theory. *Trends in Ecology and Evolution* 21:451–7.

[16] J. Hey. 1999. The neutralist, the fly and the selectionist. *Trends in Ecology and Evolution* 14:35–8; X. S. Hu, F. He, and S. P. Hubbell. 2006. Neutral theory in macroecology and population genetics. *Oikos* 113:548–56; E. J. Leigh. 2007. Neutral theory: A historical perspective. *Journal of Evolutionary Biology* 20:2075–91.

[17] M. Kimura. *The Neutral Theory of Molecular Evolution*. Cambridge University Press, 1983, p. xiv.

indeed for any related neutral processes occurring within populations. Neutral models provide a baseline from which to calculate the effects of selection: if certain colored marbles have a selective advantage, they will become more frequent and stand out in the overall distribution of colors.

But what about discontinuous, nonlinear change? Soon after Kimura published his neutral theory, biologist Robert May began to investigate the appearance of discontinuous change in ecological models. What causes a transition from linear growth to nonlinear fluctuations? As May discovered, such changes can occur with no external forcing. In an article that quickly became a seminal text in the emerging field of complexity science, May described the effects of varying the growth parameter in a simple linear model of population growth.[18] In this equation, P_t is the current population size, P_{t+1} is the population size in the next generation, and r the population's intrinsic rate of growth:

$$P_{t+1} = rP_t(1 - P_t) \qquad (1.1)$$

For small values of r, the equation is linear: an increase in the population is proportional to an increase in the growth rates. But at $r = 3.44949$, the population begins to oscillate between two values (Figure 1.1). Between 3.44949 and 3.54409, it oscillates between four values, after which slight increases in the growth rate lead to oscillations between 8, 16, 32 values, etc. When r reaches 3.56995, regular oscillations begin to be replaced by chaotic fluctuations. At these higher growth rates, tiny differences in the initial population size yield all possible final population sizes within a given range. Even more surprisingly, between 3.56995 and 3.82843 several islands of stability appear (the white "stripes" in Figure 1.1).

Online Resource: The Logistic Map

The logistic map model is available in the online resources for *Islands of Order*:

https://www.islandsoforder.com/the-logistic-map.html

Thus, merely varying the growth rate triggers linear, oscillatory, and chaotic behavior. In the language of complexity, or more specifically of nonlinear dynamics, each of these features is called a *regime*, or *attractor*.

[18] R. M. May. 1976. Simple mathematical models with very complicated dynamics. *Nature* 261:459–67.

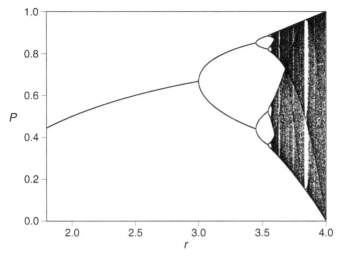

Figure 1.1. The logistic map. This phase portrait shows linear, cyclical, and chaotic behavior at different values of the intrinsic growth rate, r. Credit: Ning Ning Chung.

As long as the growth rate is less than 3.44949, the behavior is linear. But if the growth rate happens to fall in the chaotic regime, prediction is impossible, even if everything about the system is known exactly.[19]

This example allows us to make two observations. First, one need not seek very far to discover nonlinear processes. Indeed, as Stanisław Ulam famously quipped, "to speak of 'nonlinear science' is like calling zoology the study of 'nonelephant animals.'"[20] Second, simple linear processes can trigger unexpected nonlinear effects, and if more than one attractor or regime exists—that is, if the system is not a simple equilibrium—the resulting variation in dynamical behavior can easily be mistaken for noise or error. As May observed, "the very simplest nonlinear difference equations can possess an extraordinarily rich spectrum of dynamical behavior, from stable points, through cascades of stable cycles, to a regime in which the behavior (although fully deterministic) is in many respects 'chaotic,' or indistinguishable from the sample function of a random process."[21]

---

[19] Ibid.

[20] Quoted in D. Campbell, J. Crutchfield, J. Farmer, and E. Jen. 1985. Experimental mathematics: The role of computation in nonlinear science. *Communications of the Association for Computing Machinery* 28:374–84.

[21] May, Simple mathematical models.

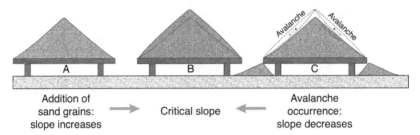

Figure 1.2. The sand pile experiment, showing critical transitions. Credit: Yves Descatoire.

Triggers for nonlinear transitions

Among the most interesting nonelephant animals are the ones that exhibit tendencies to self-organize. Stuart Kauffman, one of their discoverers, called this *order for free*. Even in the absence of selection, seemingly random local interactions can trigger the emergence of order at a higher scale.[22] An intriguing example is a behavior called *self-organized criticality* (SOC), for which the canonical example is not an equation, but an experiment often performed by toddlers at the beach.[23]

Take a flat surface, dribble grains of sand on it until it becomes a pile, and observe the occasional avalanches that occur as the sides grow steep (Figure 1.2). As the grains of sand fall, avalanches continue until the steepness of the sides remain constant. At this point, the sand pile has reached its attractor; the size of avalanches (the number of grains of sand that move) is inversely related to their frequency. That is, we see many small avalanches and few large ones. Having reached its attractor, the shape of the sandpile does not change, though it can grow larger, as long as sand flows onto it and there is enough room for the sand pile to spread.

This system has several interesting features, notably that it is self-organizing and generates a robust pattern of emergent, scale-invariant behavior (the relationship between the size and frequency of avalanches). This pattern is seen widely; for instance, the magnitude of earthquakes is inversely related to their frequency. Many social and cultural phenomena also exhibit this pattern. Self-organized criticality spontaneously generates scale-free networks, in which the degree distribution of nodes— how many connections they possess to other nodes—is inversely related

[22] In some cases, order emerges from the collective behaviors of large ensembles of smaller-scale units; in other cases, the pattern is imposed by larger-scale restraints. S. A. Levin. 1992. The problem of pattern and scale in ecology. *Ecology* 73:1943–67.

[23] P. Bak, C. Tang, and K. Wiesenfeld. 1987. Self-organized criticality: An explanation of 1/f noise. *Physical Review Letters* 59:381–4.

to their frequency. Thus, self-organized criticality is governed by a single attractor that produces a characteristic signature.

Sand piles have a single attractor. The possibility that real-world complex systems might have more than one attractor was demonstrated by the discovery of alternate stable states in Dutch lakes. For decades after the Second World War, excess fertilizer flowed into lakes in the Netherlands, providing free nutrients and triggering algae blooms. Later, the amount of fertilizer entering the lakes was reduced, but intriguingly, the lakes did not return to their original clarity. It turned out that alternate stable states (or attractors) exist in these lakes: one turbid, the other clear. In ecology, these alternate stable states or attractors are called regimes. The effects of nutrient flows depended on which regime a given lake happened to be in, so earlier studies that generalized across all lakes obscured these differences. But once the existence of alternate regimes was recognized, a simple intervention was sufficient to restore the lakes to health. Temporarily removing the fish allowed sediment to settle and zooplankton populations to increase, whereupon water clarity could be improved by reducing the amount of fertilizer flowing into the lakes.[24] The fish were then re-introduced. A take-home message is that complex systems are not necessarily symmetrical: here, as is often the case, it was easier to get into a mess than get out of it.

The comparative study of processes like this produced new theoretical insights by ecologists into the transitions between attractors. As a dynamical system approaches the boundary between alternate attractors, it will exhibit certain generic properties. These telltale signs have now been observed in many natural systems.[25] This behavior has yet to be conclusively demonstrated for social phenomena, but has triggered substantial interest due to its potential relevance for understanding critical transitions in social systems.

Complex adaptive systems

As we have just seen, complex systems are simply aggregates of interacting elements. If the elements are adaptive agents (in other words, if they exhibit purposeful or goal-seeking behavior), then they form a complex adaptive system (CAS). Complex adaptive systems are ubiquitous in the life sciences, and we are just beginning to notice them in the social world.

[24] J. L. Attayde, E. H. Van Nes, A. I. L. Araujo, et al. 2010. Omnivory by planktivores stabilizes plankton dynamics, but may either promote or reduce algal biomass. *Ecosystems* 13:410–20.

[25] M. Scheffer, J. Bascompte, W.A. Brock, et al. 2009. Early-warning signals for critical transitions. *Nature* 461:53–9.

Is a given system composed of adaptive agents, and does it exhibit emergent features that arise from their aggregate behavior? What might such emergent features look like? When do quantitative differences turn into qualitative transformations? Like the logistic equation for populations described above, even the simplest examples of complex adaptive systems can contain surprises.

To see this, we can turn the logistic equation from Figure 1.1 into an evolving complex adaptive system by adding a single environmental parameter—causing growth to be affected by some feature of the environment. The resulting model, created in 1992, helped trigger a revolution in the environmental sciences.

The model is called Daisyworld[26] and the environmental variable is temperature. Daisyworld is an imaginary planet orbiting a star like the sun and at the same orbital distance as the Earth. The surface of Daisyworld is fertile earth, sown uniformly with daisy seeds. As is true in our world, the daisies vary in color, and daisies of similar color grow together in patches. As sunshine falls on Daisyworld, the model tracks changes in the growth rate of each variety of daisy and changes in the amount of the planet's surface covered by different colored daisies.

The simplest version of this model contains only two varieties of daisies, white and black. Black daisies absorb more heat than bare earth, while white daisies reflect sunlight. Consequently, clumps of same-colored daisies create a local microclimate for themselves, slightly warmer (if they are black) and slightly cooler (if white) than the mean temperature of the planet. Both black and white daisies grow fastest, and at the same rate, when their local effective temperature (the temperature within their microclimate) is 22.5°C. They respond identically, with a decline in growth rate, as the temperature deviates from this ideal. As a result, at a given average planetary temperature, black and white daisies experience different microclimates and therefore have different growth rates.

If the daisies cover a sufficiently large area of the surface of Daisyworld, their color affects not only their own microclimate, but also the albedo or reflectance of the planet as a whole (Figure 1.3). Like our own sun, the luminosity of Daisyworld's star has gradually increased. A simulation of life on Daisyworld begins in the past with a cooler sun. This enables the black daisies to spread until they warm the planet. Later on, as the sun grows hotter, the white daisies grow faster than black ones, cooling the planet. So over the history of Daisyworld, the warming sun gradually changes the proportion of white and black daisies, creating the global phenomenon of temperature regulation: the planet's temperature is held near an optimum for—and by—the daisies.

[26] J. E. Lovelock. 1992. A numerical model for biodiversity. *Philosophical Transactions of the Royal Society B* 338:383–91.

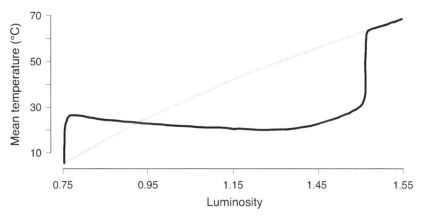

Figure 1.3. Simulated temperature regulation on Daisyworld. As the luminosity of its aging sun increases from 0.75 to 1.5 times the average value, the temperature of a bare planet would steadily rise (gray line). In contrast, the temperature of Daisyworld stabilizes close to 22.5°C when daisies are present (black line). Credit: Authors, adapted from James Lovelock's Daisyworld model.

Imagine that a team of astronauts and planners is sent to investigate Daisyworld. They would have plenty of time to study the only living things on the planet, and they would almost certainly conclude that the daisies had evolved to grow best at the normal temperature of the planet, 22.5°C. But this conclusion would invert the actual state of affairs. The daisies did not adapt to the temperature of the planet; instead they adapted the planet to suit themselves.[27] A Daisyworld without daisies would track the increase in the sun's luminance (gray line), rather than stabilizing near the ideal temperature for daisies (black line). But the role of the daisies in keeping the planet at a cozy temperature would not be obvious to the newcomers. Only when the sun's luminosity becomes too hot for the daisies to control—the abrupt transition in the black line on the right of the graph—would the daisy's former role in temperature stabilization become apparent.

Lacking this understanding, planners hoping to exploit Daisyworld's economic potential for the interstellar flower trade would fail to appreciate the possible consequences of different harvesting techniques. While selective flower harvests would cause small, probably unnoticeable tremors in planetary temperature, clear-cutting large contiguous patches of daisies would create momentary changes in the planet's albedo that

[27] P. T. Saunders. 1994. Evolution without natural selection: Further implications of the Daisyworld parable. *Journal of Theoretical Biology* 166:365–73.

could quickly become permanent, causing temperature regulation to fail and daisy populations to crash. Something quite like this happened during the 1970s Green Revolution on the Indonesian island of Bali, as we will see in chapter 5.

The Daisyworld model soon became a canonical example of a self-organizing, self-regulating environmental system. As an example of a complex adaptive system, it has several interesting features. Unlike sand piles, this model is driven by a process of adaptation. And the biology of adaptation is as simple as its creator, James Lovelock, could possibly make it. The model shows how small-scale local adaptations can trigger an emergent global structure (temperature regulation at the planetary scale). And it also shows why such global structures can easily fade from view, becoming noticeable only when the system as a whole has been pushed past its limits.

Online Resource: Daisyworld

The Daisyworld model is available in the online resources for *Islands of Order*:

https://www.islandsoforder.com/daisyworld.html

Discovering islands of order

The doomed flower markets of Daisyworld conclude this overview of models of change, which we build upon in the case studies that follow. Older models of stadial change and stable equilibria remain of interest, but we suggest that they are best treated as special cases.

Where do we go from here? A broader conceptual framework is needed to detect complex emergent phenomena. Such a framework does not yet exist for the social sciences, but an obvious way forward is to take advantage of two existing frameworks that are commonly used in complexity research and offer complementary insights. The first is attractor basins from physics, the second adaptive landscapes from evolutionary biology. Because we will use both of these ideas in future chapters, we offer a brief introduction to them here.

Phase portraits and basins of attraction

Phase portraits offer a simple and intuitive snapshot of the behavior of dynamical, evolving systems. We have already encountered an example of a phase portrait in Figure 1.1, the logistic map. For convenience, this

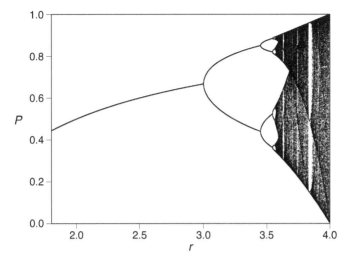

Figure 1.4. The logistic map, revisited. Credit: Ning Ning Chung.

figure is reprinted above (Figure 1.4). The way to read it as a phase portrait is to mentally slide along the horizontal axis, tracking increases in r (the growth rate) and glancing up to see how P (the population size) changes. At different values of r, the population undergoes stable, oscillatory, complex, and chaotic behavior. Each of these patterns is an attractor. The span of r values that trigger a particular pattern is the *basin of attraction* for that attractor. (The analogy being a geographical drainage basin, where rain falling on some area inexorably flows into the region's main river, here analogous to the attractor.) In Figure 1.4, the largest basin is for stable (linear) growth, which extends to $r = 3$. Above 3, there is a new basin of attraction for oscillatory dynamics, where the population oscillates between two values, which appear as "branches" on the figure. A third attractor appears around 3.4, where the population oscillates between four values (the "bubbles"). This basin is smaller, confined to the interval, $3.44949 < r < 3.54409$, after which another basin appears. Further increases in r explore many tiny basins, associated with oscillatory, complex, and chaotic attractors, and even a (brief) return to stable growth (the white "stripes"). Thus, in the region $r > 3.4$, this phase portrait is characterized by an abundance of many tiny attractors.

With this example, we draw your attention to the importance of basins of attraction. To create a phase portrait, the key question is not only the nature of the attractors, but the regions of the phase plane that drain into them: their basins of attraction. In general, it is rare to find systems that comprise a single basin draining to a point attractor (in other words, standard social science equilibrium models). But the simplicity of these models makes them ideal null or neutral baselines. An interesting example is Motoo Kimura's model of neutral equilibrium (described above), which describes patterns of change driven by chance alone. This system has a single attractor; the larger the population, the slower the approach to the attractor. Kimura's neutral model provides the theoretical framework for our analysis of male dominance in chapter 3.

Daisyworld is a slightly more complex model. It has three basins of attraction, which depend on the amount of sunlight reaching the planet.[28] This model is the starting point for our analysis of Balinese water temple networks in chapter 5. The Bali model is based on a simple dynamical relationship, much like Daisyworld. But what if the data are noisy and neither the attractors nor their basins are readily apparent? We will pick up that question in chapter 7, where we consider how to discover basins of attraction in noisy data from a social survey.

Definitions

Complex systems research uses a number of concepts and terms that you may not have seen before. Here are some basic explanations of key ideas we have encountered so far.

Emergent properties are a characteristic of systems in which you cannot predict outcomes by observing the actions of an individual, but only when you see many individuals interacting together. This is the opposite of reductionist science, which aims to reduce a system to its smallest parts. With an emergent property, seemingly random local interactions between individuals can often trigger the *emergence* of order at higher scales. We show some examples in later chapters.

A *phase space* is a mathematical construct that represents every possible state in a system, with each state having a unique point in the phase space. For a dynamical system with just two variables, like P and r in the logistic map, you can imagine the phase space as a two-dimensional plot with P and r on the axes. (Strictly, a phase space in two dimensions is usually called a *phase plane*.) If a system has three variables, the phase space is three-dimensional. If it has ten variables, the phase space is ten-dimensional. Mathematically, all of

---

[28] Attractor basins can be calculated and plotted for discrete dynamical systems in 1, 2, or 3 dimensions using Discrete Dynamics Lab, http://www.ddlab.com.

these cases work exactly the same way. It is just harder to visualize examples with more than three dimensions.

A plot that shows the outcome of some set of initial conditions in a given phase space is called a *phase portrait*. For example, Figure 1.4 shows the phase portrait for the two-dimensional dynamical system of the logistic map.

In many dynamical systems, there is a part of the phase space where initial conditions inevitably evolve to a particular final state. That final point is called an *attractor* and the area around it is called a *basin of attraction*. An attractor is a set of states that neighboring states in a given basin of attraction asymptotically approach during the course of dynamic evolution. Think of a water analogy: rain falling within a watershed inevitably flows into the region's major river. The watershed is the basin of attraction and the river is the attractor.

You are already familiar with attractors, even if you do not know it. In many cases (but not always), data points that are normally distributed—like the lengths of leaves on a tree or the volume of a certain pot type in an archaeological assemblage—actually belong to a dynamical system with just one attractor. The peak of the bell curve is usually the attractor and the curve around it the basin of attraction. The dynamical systems in this book are novel in that they usually have multiple attractors, each with its own basin of attraction. The presence of—and interaction between—multiple attractors leads to more complex system dynamics. Describing those behaviors is a key purpose of this book.

Adaptive landscapes

The concept of adaptive landscapes[29] (also called *fitness landscapes*) was proposed by biologist Sewall Wright in 1932, and is now probably the most common metaphor used in evolutionary genetics.[30] Unlike

---

[29] S. Wright. 1932. The roles of mutation, inbreeding, crossbreeding and selection in evolution. *Proceedings of the Sixth International Congress of Genetics* 1:356–66.

[30] "Adaptive landscape is probably the most common metaphor in evolutionary genetic[s]." D. J. Futuyma. *Evolutionary Biology*, Sinauer Associates, 1998, p. 403. For discussion of the limitations of the adaptive landscape concept in biology, see P.A.P. Moran. 1964. On the non-existence of adaptive topographies. *Annals of Human Genetics* 27:383–93; G. Gilchrist and J. Kingsolver. 2001. Is optimality over the hill?

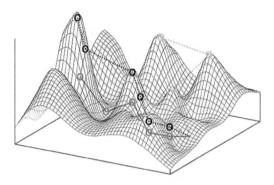

Figure 1.5. An example of a fitness landscape in two dimensions. The lines represent alternative mutational paths to reach different peaks in the landscape. Note that environmental and social change means that the landscape itself is not static, as might be the initial impression from this figure, but instead changes in dynamic ways over time. Credit: Randy Olson, Wikimedia Commons, CC BY-SA 3.0.

phase portraits, adaptive landscapes do not lend themselves to rigorous mathematical analysis. Instead they provide a way to visualize trajectories of change in evolving complex adaptive systems. We combine the concept of adaptive landscapes with phase portraits in chapter 7, where we investigate how social systems can move between basins of attraction.

The idea of an adaptive landscape is intuitively simple: imagine a collection of evolving agents—distinct entities such as organisms, people, or strategies—on a surface, where their height in this space reflects the relative fitness of each agent (Figure 1.5). An adaptive process, if one is present, will move the population from valleys to peaks. The fittest organisms cluster around the highest peaks, while the lowest fitness is represented by deep valleys.

These peaks can take different forms. The simplest is the "Mount Fuji" landscape with a single fitness peak (a Gaussian distribution of fitness). In contrast, if all the fitnesses are identical, the result is a flat fitness landscape. Here, there is no variation in fitness, so natural selection has nothing to work with. Between these extremes, more irregular distributions of fitness produce a rugged fitness landscape, with peaks of varying height. Because the rate of reproduction of an organism or agent is determined by its fitness, selection will cause an evolving population

The fitness landscapes of idealized organisms. In S. Orzack and E. Sober, eds. *Adaptationism and Optimality*. Cambridge University Press, p. 219–41; M. Pigliucci and J. Kaplan. *Making Sense of Evolution: The Conceptual Foundations of Evolutionary Biology*. University of Chicago Press, 2006; B. Calcott. 2008. Assessing the fitness landscape revolution. *Biology and Philosophy* 23:639–57.

to climb uphill in the fitness landscape over time, until it reaches a local optimum. If more than one fitness peak exists, populations can get stuck on lower peaks, never reaching the higher ones.

Greater realism can be introduced into fitness landscapes in several ways. Instead of assigning a permanent fitness to each organism, fitnesses can be allowed to vary. For example, the fitness of organism (or strategy or agent) A may depend in part on organisms B and C. The landscape itself can also change shape as the populations explore it. This concept—dynamic adaptive landscapes—plays a prominent role in evolutionary game theory, in which the evolving entities are strategies and their payoffs (fitness) depend on their relative frequency.

In this book, we will follow the advice of a philosopher of science, Peter Godfrey-Smith, and use adaptive landscapes to help decide what kind of model is best suited to a given question. For Godfrey-Smith, the key question is scale. At very small scales of space and time, where all agents are visible as points on the landscape, movement on the landscape may be dominated by neutral drift rather than selection. After all, mutations are rare and most new mutations do not provide a fitness advantage. Instead, neutral mutations tend to accumulate. At this scale, natural selection is just one factor among many, and will rarely be dominant. So it makes sense to begin with a neutral model, and then look carefully for evidence of selection or other kinds of non-neutral change.

At longer time spans, evolutionary game theory starts to become relevant. Here, as Godfrey-Smith points out, "[t]he fine details of population movements on the landscape are washed out and replaced by idealized strategies, whose competition drives a selection process. Paleontologists often zoom out even further, considering observed forms in contrast to a broad range of hypothetical (unobserved) alternative types. At this coarsest grain of analysis, selection again recedes in perceived importance, as the large set of conceivable alternatives highlights the great importance of historical contingency in producing observed forms."[31]

Our most zoomed-out case study, the colonization of the Pacific, takes us back over 150 generations, just brushing the Pleistocene. And sure enough, at this scale there is unmistakable evidence of selection. At the other extreme, decisions about cooperation are nearly simultaneous and can appear to be nearly random. We agree with Godfrey-Smith that the question of scale is relevant to any theory of change in an evolving population, which makes adaptive landscapes a very useful metaphor. So we have taken his advice, and begin each of our case studies by posing a question or questions, and then zooming in to the relevant scale.

[31] J. F. Wilkins and P. Godfrey-Smith. 2009. Adaptationism and the adaptive landscape. *Biology and Philosophy* 24:199–214.

Conclusion

In one of the foundational articles that launched complexity studies, physicist Philip Anderson rephrased Karl Marx's observation that quantitative differences become qualitative differences. "More is Different," Anderson observes, because at each new level of complexity entirely new properties appear.[32] The way to discover these emergent properties is by tracing patterns of interaction among the elements of a given system. The phenomenon of emergence is common to all of the examples we have considered in this chapter, and will continue to be relevant in each of the case studies in this book. But until recently the mathematical toolkit for analyzing adaptive change was not well suited to discovering emergence or other properties of out-of-equilibrium dynamical systems. As recently as 1990, philosopher of science Karl Popper argued that social scientists who wish to take advantage of mathematics have the choice of only two approaches.[33] The first is essentially Newtonian and is best represented by general equilibrium theories (for example, in economics). Such theories take the form of systems of differential equations describing the behavior of homogeneous social actors. Change occurs as a result of perturbations and leads from one equilibrium state to another. The second type of theory is statistical. If one cannot write the equations to define a dynamical system, it may yet be possible to observe statistical regularities in social phenomena. Both approaches have obvious weaknesses: the assumption of equilibrium is forced by the mathematics, not by observations of social behavior; and sifting for patterns with descriptive statistics is at best an indirect method for discovering causal or developmental relationships.

We are hardly the first to comment on these limitations. In fact, they were the central issue in what is generally reckoned to be the most influential debate about the methodological foundations of social science of the last century, the "Positivismusstreit" or "Positivist Dispute"[34] between Popper and the social theorists of the Frankfurt School from 1961 to 1963. Popper argued that progress in the social sciences was achievable only by the use of mathematics to falsify hypotheses. In response, Theodor Adorno observed that descriptive statistics provide no explanation for qualitative change, or what we would now call emergence: "only through what it is not will it disclose itself as it is."[35] This led

---

[32] P. W. Anderson. 1972. More is different. *Science* 177:393–6.

[33] K. Popper. 1990. *A World of Propensities*. Thoemmes Press, 1990, pp. 18–19.

[34] T. W. Adorno, H. Albert, R. Dahrendorf, et al. *The Positivist Dispute in German Sociology*, transl. Glyn Adey and David Frisby, Heinemann, 1976.

[35] Ibid., 296.

Adorno to a critique of descriptive statistics as the primary tool for social inquiry. He observed that "a social science that is both atomistic, and ascends through classification from the atoms to generalities, is the Medusan mirror to a society which is both atomized and organized according to abstract classificatory principles. ... " Adorno's point was that a purely descriptive, statistical analysis of society at a given historical moment is just "scientific mirroring" that "remains a mere duplication." To break the seal of reification on the existing social order, it would be necessary to go beyond descriptive statistics or equilibrium models to explore historical contingency. However, the mathematical tools that might enable this kind of investigation did not yet exist, and the Positivist Dispute ended in a stalemate.

Still, the question of historical contingency would not go away. In the 1980s, sociologist Anthony Giddens developed an influential theory of *structuration*, arguing that human social activities, like some self-reproducing items in nature, are recursive. That is to say, they are not brought into being by social actors, but continually recreated by them via the very means whereby they express themselves as actors. In and through their activities, agents reproduce the conditions that make these activities possible.[36] But Gidden's theory was pitched at a very general level, a description of the human condition rather than a methodology for investigating specific processes of change.

The theoretical landscape looks very different today. One important change since the 1980s has been the flourishing of computational modeling. But the availability of more powerful tools for statistical analysis is only part of the story. Our subject in this chapter has been the implications of the discovery of Ulam's nonelephant animals: attractors in nonlinear systems. As Robert May showed with his logistic map, they are not hard to find, once we learn to recognize them, and their discovery has opened up new vistas in physics and biology. As relative latecomers to this perspective, social scientists are in a position to benefit from several decades of theoretical work, including a substantial body of elegant mathematical tools.

But how to make use of these ideas? In the chapters that follow, we offer some suggestions.

---

[36] A. Giddens. *The Constitution of Society: Outline of the Theory of Structuration.* University of California Press, 1984.

CHAPTER 2

~~~~~~

Discovering Austronesia

He wāhine, he whenua, ka ngaro te tangata.[1]
—Traditional saying (whakataukī) of New Zealand Māori

Introduction

Since our species first appeared, our ancestors have mostly lived in small groups: networks of nomadic or settled communities that were often widely dispersed across the landscape, in striking contrast to the large, urbanized and highly interconnected societies that have emerged within the last few thousand years. But small traditional societies that resemble prehistoric settings have been replaced in most regions, including large parts of Europe, Africa, Asia, and the Americas, by frequent post-Neolithic population movements. Further population restructuring has been driven by the actions of modern states during the historic era (for instance, see Stephen Leslie and colleagues for a history of these processes in the British Isles).[2] As a result, where small, seemingly traditional communities exist, their histories may extend only into the recent past.

An intriguing exception is Iceland, where parish records include the births, marriages, and deaths of the entire population from the time the island was first colonized in the ninth century. These records have enabled geneticists to track the transmission of genes related to health over the centuries. Knowing each person's complete genealogy on both the father's and mother's side makes it possible to follow specific genes from one generation to the next. In this way, geneticists discovered more than 8,000 individuals with so-called *knockouts*—rare genetic mutations that disable a gene—and traced the consequences of these knockouts in the evolving population. A similar study in neighboring Scandinavia would be impossible, because there have been many population movements, and genealogies would be partial at best. But despite its advantages, the Icelandic study is limited in scope, because the population is small (325,000 people) and homogeneous, with relatively little genetic variation.

---

[1] "For women and land, men will perish."
[2] S. Leslie, B. Winney, G. Hellenthal, et al. 2007. The fine-scale genetic structure of the British population. *Nature* 519:309–14.

The islands of the more remote regions of Island Southeast Asia and Oceania offer the possibility to undertake more ambitious studies. The history of Iceland as an isolated population goes back a thousand years. The islands of the Austronesian world were colonized much earlier, there are a lot more of them, and their environments are more varied. But the most interesting difference relates to their origins. Iceland was colonized just once, but two great migrations populated the Pacific islands. One, about 50,000 years ago, took modern humans all the way from the Asian mainland to Australia. The second, which commenced about 4,000 years ago, spread across half the globe and ended in the last great human expansion into pristine environments, the settlement of Oceania.[3] By now, one might expect that later population movements would have erased most traces of ancient migrations. Instead, as we will see in this chapter, like Iceland, many Austronesian populations remained quite isolated after their initial colonization. Consequently, using genetic information, it is possible to reconstruct detailed histories of both migrations, including what happened on the islands where the new arrivals met the descendants of the original Pleistocene migrations. These dated chronologies of population movements and mingling can be interpreted with insights from anthropology and linguistics. Thanks to the vast distances between Pacific islands, they offer perhaps the most comprehensive record of the evolution of social structure and language in what paleontologists call *deep time*.

The first part of this chapter sets the stage with an account of the history of migrations into the islands, beginning with the earliest journeys out of Africa. The second part describes how we sought to make sense of the two biggest surprises in the genetic and linguistic data, and the third part explains the analytical models we developed. The time scale for the models runs from tens to hundreds of generations, and the spatial scale is Island Southeast Asia and the Pacific.

Dubois' remarkable discovery

Eugène François Thomas Dubois (1858–1940) grew up in Holland. As a schoolboy he collected fossils from limestone mines around Limburg, and attended lectures on Charles Darwin's new theory of evolution. Later he trained as a medical doctor and took a postgraduate degree in comparative anatomy. Fascinated by the theories of Ernst Haeckel on human evolution, and inspired by the discovery of the first Neanderthal skeletons

---

[3] We refer to the migration of Austronesian-speaking peoples, which certainly dates to that period. Other population movements are not part of the story we tell here.

in the nearby Belgian town of Spy, Dubois decided to search for fossil evidence for the transition from apes to humans. While Darwin argued that this probably occurred in Africa, an alternative hypothesis was proposed by Alfred Russel Wallace, who shares credit with Darwin for the discovery of natural selection. Wallace was a British naturalist who explored the Malay archipelago from 1854 to 1862, amassing a collection of more than 126,000 specimens. In 1869, Wallace published *The Malay Archipelago*, which became one of the most popular books of scientific exploration of the nineteenth century, and has never been out of print. Wallace argued that humans are more closely related to gibbons and orangutans than African apes, and so must have evolved in the Malay archipelago. Dubois decided to seek proof for this idea. In 1887 he joined the Dutch army as a military surgeon and arranged to be posted to the Dutch East Indies (now Indonesia).

In his spare time, Dubois began to excavate caves in Sumatra, and was soon rewarded with the discovery of Pleistocene fossils of extinct mammals. Impressed, the colonial government relieved him of his military duties and assigned teams of engineers and convicts to assist him in his search for human ancestors. In August 1891, Dubois discovered the first of many fossils along the Solo river in Sangiran, East Java. A skull cap with a cranial capacity of 900 cc (cubic centimeters) and a femur (thigh bone) suggesting upright bipedalism led Dubois to propose that they belonged to a transitional form between apes and humans which he named *Pithecanthropus erectus* ("upright ape-man"). Today, fossil bones of this species (now classified as *Homo erectus*) continue to be unearthed at Sangiran, which is now a UNESCO World Heritage site.

Dubois' remarkable discovery of the upright ape-man triggered a worldwide search for ancient bones. However, as it turned out, Darwin was right: our closest relatives are not gibbons and orangutans, but African apes. Consequently, *Homo erectus* must have migrated from Africa across Asia to the islands of Indonesia. Later discoveries made it possible to trace this route. Until recently, anthropologists believed that the descendants of *Homo erectus* became extinct in the region long before the appearance of modern humans. Then in 2003 archaeologists excavating a cave on the Indonesian island of Flores discovered the remains of a very small archaic hominin,[4] which they named *Homo floresiensis* ("Flores Man"; nicknamed the "hobbit"). The skeletal material was found to date from approximately 100,000 to 50,000 years ago, a period that overlaps with modern humans. The origin of the

---

[4] Members of the human clade, that is, the Hominini, including *Homo* and those species of the australopithecines that arose after the split from the chimpanzees, are called *hominins*. Not all hominins are ancestral to our species.

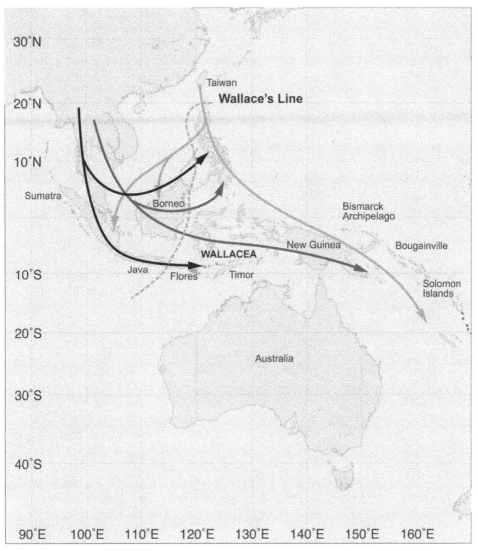

Figure 2.1. Migration routes from Asia into the Pacific. During the Pleistocene, which lasted until about 12,000 years ago, sea levels were much lower. Australia and New Guinea formed a single continent called Sahul, and most of the islands of what is now Indonesia were connected to Asia in a land mass called Sunda. Sunda and Sahul remained separated during the Pleistocene by deep waters in a region called Wallacea, named for the British naturalist Alfred Russel Wallace. Around 50,000 years ago, humans migrated across the Wallace Line (dashed line) and colonized New Guinea and Australia (black arrows). Much later, about 4,000 years ago, a second wave of migrations brought Asian farmers and fishermen across the Wallace Line and out into the Pacific (gray arrows). Credit: Yves Descatoire.

hobbits remains a mystery, but one hypothesis holds that they are the descendants of *Homo erectus*. Their small stature could be the result of island dwarfism, which has been observed for mammals on many islands, but never before for humans. Whatever their origins, as soon as the bones were dated, archaeologists began to speculate: could modern humans and hobbits have met on Flores, and perhaps coexisted? When did modern humans arrive in the islands?

The first migration of modern humans

Homo erectus made their way to the islands in the middle of the Pleistocene, an age of alternating cold and warm spells that lasted from about 2,588,000 to 11,700 years ago. During the cold intervals, sea levels dropped by one hundred or more meters, and the large islands of Indonesia (Borneo, Sumatra, Java) (Figure 2.1) became connected to the Asian mainland, forming a continental land mass called Sunda, crisscrossed by vast now-submerged rivers. This would have made it easy for *Homo erectus* to migrate south, and the cold climate of the north would have made equatorial Sunda an attractive destination.

The lower sea levels of the Pleistocene also joined New Guinea, Australia, and Tasmania into a single continent known as Sahul. But even when sea levels were at their lowest point, Sahul and Sunda were separated by deep-water straits. Today this region of separation is called Wallacea, named for the British naturalist. Wallacea is a major biogeographical frontier that few land mammals were able to cross during the Pleistocene. *Homo erectus* got as far as the eastern edge of the Sunda shelf, but no further. The ancestors of the hobbits managed to cross the Pleistocene seas to reach the island of Flores in Wallacea. We now have hints that archaic humans were also able to reach Sahul.

The sea journey across Wallacea to New Guinea and Australia would have required boats, and would not have been easy for hunter-gatherers. Yet archaeological findings indicate that modern humans reached the interior of Australia at least by 50,000 years ago.[5] Many prehistorians find this date to be surprisingly early, because it is not much later than the estimated date of the first human forays out of Africa. More surprisingly, the presence of stone tools indicates that humans also reached the remote

[5] J. F. O'Connell and J. Allen. 2004. Dating the colonization of Sahul (Pleistocene Australia–New Guinea): A review of recent research. *Journal of Archaeological Science* 31:835–53; J. F. O'Connell, J. Allen, M.A.J. Williams, et al. 2018. When did *Homo sapiens* first reach Southeast Asia and Sahul? *Proceedings of the National Academy of Sciences USA* 115:8482–90.

New Guinea highlands by roughly the same time. At an elevation of ~2,000 meters, even today temperatures in the highlands drop to below freezing at night, and it would have been colder 50,000 years ago.[6]

Thus, beginning around 50,000 years ago, both Australia and New Guinea were populated by modern humans with stone tools, clothing, and technology that enabled them to not only hunt and gather, but clear small patches of forest to let in sunlight and grow useful plants. Genetics as well as archaeology indicates that thenceforth these populations remained largely isolated from each other and from the Asian mainland for the next 45,000 years. The genetic evidence suggests that only a few people from the first migration traveled all the way to New Guinea and Australia. Many more remained in Sunda and Wallacea. Quite possibly some of them became acquainted with hobbits.

The second migration: Austronesians

Beginning around 12,000 years ago, the end of the Pleistocene brought rising seas that flooded the ancient continent of Sunda, creating the Java and South China Seas and the thousands of islands that make up the Malay archipelago. Sunda's inhabitants would have been compelled to retreat from the coasts into the interior of the islands. These small populations of hunter-gatherers were largely cut off from the Asian mainland until about 4,000 years ago, when sea-going colonists began to arrive. The new arrivals were Neolithic peoples from Asia, who brought with them new technologies for horticulture, fishing, and a social system based on settled communities rather than nomadic foraging.

At about the same time or perhaps even earlier, a second Neolithic transition from foraging to farming got under way in the highlands of Papua New Guinea. Taro and yams were cultivated by 10,200 BP (years before present), later followed by bananas, sago, and sugar cane.[7] Around 3000

---

[6] Data from the New Guinea Highlands (at an elevation of ~2,000 meters) demonstrate the exploitation of the endemic nut *Pandanus* and yams in archaeological sites dated to 49,000 to 36,000 years ago, which are among the oldest human sites in this region. The sites also contain stone tools thought to be used to remove trees, which suggests that the early inhabitants cleared forest patches to promote the growth of useful plants. G. R. Summerhayes, M. Leavesley, A. Fairbairn, et al. 2010. Human adaptation and plant use in Highland New Guinea 49,000 to 44,000 years ago. *Science* 330:78–81.

[7] R. Fullagar, J. Field, T. Denham, and C. Lentfer. 2006. Early and mid Holocene tool-use and processing of taro (*Colocasia esculenta*), yam (*Dioscorea* sp.) and other plants at Kuk Swamp in the highlands of Papua New Guinea. *Journal of Archaeological Science* 33:595–614.

BP, the two Neolithic cultures, Papuan and Asian, came into contact when the sea-going Austronesians reached Wallacea and the northern coast of New Guinea. The results of that contact are the subject of this chapter. Three sources of information provide complementary insights: archaeology, language, and DNA.

The archaeological record is relatively sparse, which is not entirely the fault of the archaeologists. Nearly half the Sunda shelf became submerged at the close of the Pleistocene a little over 10,000 years ago, concealing any evidence of human habitation along the coasts and river valleys. As sea levels rose, the island realm of Wallacea expanded to the east, while New Guinea separated from Australia. The earliest dated Neolithic sites in Papua New Guinea are found in the highlands. The first cultivated crop was probably taro, and taro with the same DNA fingerprint as around these early sites is found both east of New Guinea in the Solomon Islands and west in the islands of Wallacea. This raises an as yet unanswered question: was the spread of taro and the other crops caused by migrations of Neolithic farmers from the highlands of Papua New Guinea?[8] Or were the crops (and the knowledge of how to grow them) dispersed along trade routes? A strong hint that migrations took place is provided by the geographic distribution of Trans–New Guinean languages, which closely matches the distribution of taro.

In Asia, the Neolithic expansion did not originate in the islands, but in mainland Southeast Asia and southern China. Rice was domesticated in China by about 9,000 years ago; soon after that, domesticated chickens began to appear in archaeological sites in China and mainland Southeast Asia. Pigs were apparently domesticated twice, once in Anatolia and once in China's Mekong valley. Beginning around 5,000 years ago, these and other domesticated species started to appear in Island Southeast Asia. Their dispersal routes can often be traced. For example, pigs are a traditional, highly valued resource in New Guinea. But all native mammals in New Guinea and Australia are marsupials, whereas pigs are placental mammals. Pigs have little aptitude for swimming, so the first pigs to arrive in New Guinea must have been passengers in sailing canoes. Evidence from pig mtDNA points to multiple distinct migrations both

---

[8] Although only the east New Guinea highlands have yielded clear evidence for an autochthonous development of agriculture, indigenous agriculture may have been practiced more widely in this region than we currently have archaeological evidence for, an argument made from genetic evidence. T. P. Denham, S. G. Haberle, C. Lentfer, et al. 2003. Origins of agriculture at Kuk Swamp in the Highlands of New Guinea. *Science* 301:189–93; S. Mona, M. Tommaseo-Ponzetta, S. Brauer, et al. 2007. Patterns of Y-chromosome diversity intersect with the Trans–New Guinea hypothesis. *Molecular Biology and Evolution* 24:2546–55.

eastward out of Southeast Asia and within Wallacea.[9] The dispersal of other domesticates can be traced with the same method of DNA fingerprints. Rats, for example, were carried as far as Easter Island by colonists, and left their bones (and DNA) in archaeological sites.[10]

Using these techniques—archaeology, DNA fingerprints, and the geographic distribution of language families—prehistorians can reconstruct migration routes and estimate when they occurred. This has been a very active area of research for decades, largely focused on discovering the colonization routes taken by the Austronesians into the Pacific. One question has been particularly controversial: the geographic origins of the peoples who spread the Austronesian language family across the Pacific. The Austronesian languages are thought to have originated in Taiwan, and both genetic and archaeological data support the hypothesis that Taiwan was the Austronesian homeland. However, other evidence has been used to challenge the "out-of-Taiwan" hypothesis. It has been argued that the Austronesian language could have originated elsewhere, and that genetic data may point to multiple post-Pleistocene migrations from Asia into the islands. New genetic data that bear on the question of the Austronesian homeland continue to be published, and as of this writing the question remains open. But fortunately we can set it aside for the purposes of this chapter. Our story begins later, in the era when the Austronesian migrations into the islands of Wallacea brought them into contact with the descendants of the earlier migrations.

Surprises in the data

In 2010 we discovered that one of the greatest geographic barriers to the flow of genes between human populations, equivalent to the Himalayas or the Sahara, occurs in Wallacea in the midst of a continuous chain of islands that form the southern arc of the Indonesian archipelago.[11] As explained above, Wallacea is the frontier zone separating mainland Asia

[9] P. Bellwood and P. White 2005. Domesticated pigs in eastern Indonesia. *Science* 309:381; G. Larson, K. Dobney, U. Albarella, et al. 2005. Worldwide phylogeography of wild boar reveals multiple centers of pig domestication. *Science* 307:1618–21; J. K. Lum, J. K. McIntyre, D. L. Greger, et al. 2006. Recent Southeast Asian domestication and Lapita dispersal of sacred male pseudohermaphroditic "tuskers" and hairless pigs of Vanuatu. *Proceedings of the National Academy of Sciences USA* 103:17190–5.

[10] S. S. Barnes, E. Matisoo-Smith, and T. L. Hunt. 2006. Ancient DNA of the Pacific rat (*Rattus exulans*) from Rapa Nui (Easter Island). *Journal of Archaeological Science* 33:1536–40.

[11] Analysis of molecular variance reveals one of the highest levels of between-group variance yet reported for human Y chromosome data (e.g., $\Phi_{ST} = 0.47$).

from Australasia. It was created by deep ocean straits separating the Asian and Australian continental shelves. For several million years, the straits were a barrier to migration by land mammals. But they are easily crossed by humans in small boats, and the ancestors of the Papuans did so more than 50,000 years ago. Why, then, did the genetic barrier occur? In the absence of geographic barriers to migration, the genetic boundary must have emerged from social behavior.

We discovered the barrier while investigating the genetic makeup of the islands along the southern arc. At the western edge, the islanders have mostly Asian ancestry, while to the east the inhabitants possess mostly Papuan DNA dating from the earlier migration. To gain insight into the genetic barrier in Wallacea, we identified 37 genetic markers that distinguish between Asian and Papuan ancestry, because they are commonly found in only Asian or Papuan populations.[12]

Some of the markers we chose came from X chromosomes, which spend two-thirds more time in women than in men. The X chromosomes revealed an intriguing sex bias in the data, which is shown in Figure 2.2. Markers from the X chromosomes (shown in black triangles) retain more Asian ancestry than the others. In other words, while the barrier exists for all 37 genetic markers, proportionately more of the X chromosome markers made it across the Wallace Line. This suggests that Asian women found it easier to cross the barrier than Asian men.

The linguistic data also contained a surprise. At the eastern end of the arc, today the islanders speak Papuan[13] languages. At the western end of the arc, the islanders mostly have Asian ancestry. Their ancestors were Austronesian-speaking farmers and fishermen, who arrived in Wallacea less than 5,000 years ago. As they made their way along the island chains, the Austronesians would have encountered descendants of the earlier Paleolithic migration, who must have spoken their own languages. But today those languages are gone, replaced by the great tide of Austronesian languages that came to be spoken along the entire arc

[12] Ancestry informative markers (AIMs) used in this study consist of 37 genetic markers (SNPs) that distinguish between Asian and Papuan ancestry, because they are commonly found in only Asian or Papuan populations. Thirty-seven AIMs were genotyped in the largest panel of Island Southeast Asia samples studied at that time: 1,430 individuals from 60 populations, ranging from mainland East Asia to Melanesia (Figure 2.2). Consistent with the evidence for sex-biased admixture from the earlier mtDNA and Y chromosome studies, mean rates of Asian admixture are higher on the X chromosome than on the autosomes.

[13] The Papuan languages are the non-Austronesian (and non-Australian) languages spoken on the western Pacific island of New Guinea and neighboring islands by around four million people. This is a strictly geographical grouping of languages and does not imply a shared linguistic relationship.

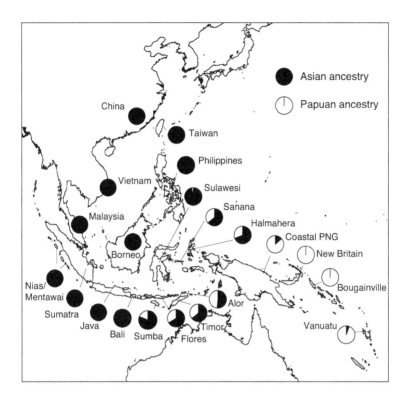

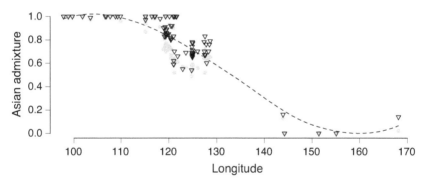

Figure 2.2. Admixture rates across Southeast Asia and Oceania. (*Upper*) Pie charts show mean regional admixture rates averaged across the autosomes and X chromosome (Asian ancestry in black; Papuan ancestry in white). (*Lower*) Change in Asian ancestry proportions calculated from all genetic markers combined (dashed line). Asian admixture estimated from autosomal and X chromosomal SNPs are represented by gray circles and black triangles, respectively. Note the decline in Asian ancestry beginning in eastern Indonesia, as well as preferential retention of Asian ancestry on the X chromosome (triangles) versus the autosomes (circles). M. P. Cox, T. M. Karafet, J. S. Lansing, et al. 2010. Autosomal and X-linked single nucleotide polymorphisms reveal a steep Asian-Melanesian ancestry cline in eastern Indonesia and a sex bias in admixture rates. *Proceedings of the Royal Society B* 277:1589–96. Credit: Authors.

of Indonesian islands, around the coast of New Guinea, and across the Pacific. Hence, the discovery of the barrier provoked three questions:

1. What caused the genetic barrier?
2. Why was it easier for women to cross?
3. When the Austronesian-speaking colonists arrived, the islands of Wallacea were already inhabited by people speaking their own Papuan languages. Why were those languages replaced by Austronesian languages?

The toolkits: Population genetics and kinship

To pursue these questions, we need to supplement the prehistorian's usual methods with some additional tools. DNA fingerprints like those used to track the movements of pigs and rats are informative, but a great deal more can be learned from genetics (Figure 2.3), especially in combination with linguistic data. Both languages and DNA evolve in predictable branching patterns, which can be analyzed both separately and in combination. Moreover, there are several ways in which different types of DNA are transmitted, each of which provides specific information about social behavior. For example, we inherit our mitochondrial DNA only from our mothers (it is not subject to sexual recombination). Some noncoding parts of this molecule undergo rapid mutation; these are called *hypervariable regions* and can be used as markers to estimate the relatedness of individuals. Comparing any two people, the less variation in their hypervariable regions, the more closely they are related. At one extreme, individuals who share a very recent common female ancestor will carry the same mitochondrial haplotype (i.e., set of neutral mitochondrial DNA markers). The rate at which these markers change is predictable. Thus, mitochondrial DNA (abbreviated mtDNA) can be used to build phylogenetic trees tracing matrilineal descent and estimate the dates at which new branches (and thus haplotypes) appeared. An analogous approach can be used to trace shared patrilineal descent, using noncoding regions on the nonrecombining portion of the paternally inherited Y chromosome (abbreviated NRY). Comparing two men, the more similar their noncoding Y chromosome sequences, the closer their patrilineal relatedness (Figure 2.4). These trees too can be dated with molecular clocks.

A third genetic system that provides information about ancestry is the X chromosome. Girls inherit two copies, one from their mother and one from their father. Boys inherit one from their mother, but they inherit a Y chromosome (instead of an X) from their father. Consequently, over time X chromosomes spend two-thirds as much time

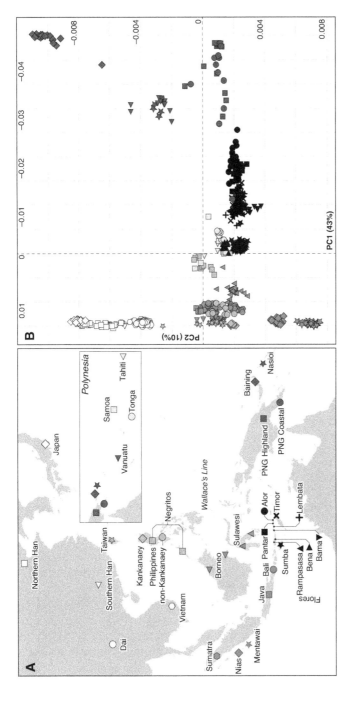

Figure 2.3. Study locations across Island Southeast Asia. *A.* Map of selected study sites. *B.* Principal components analysis (PCA) plot of genetic similarity between regional populations from genome-wide genetic data. Note the unusual separation of populations into distinct clusters, which approximately mirror their geographic locations. Credit: Georgi Hudjashov.

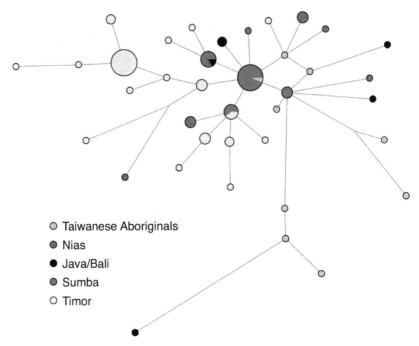

Figure 2.4. A typical genetic network showing relationships between individuals in Island Southeast Asia who belong to the Austronesian-associated Y chromosome haplogroup, O-M110. Note the haplotypes shared between islanders on Taiwan and Sumba a distance of 3,700 km, indicating shared common ancestry inherited from their fathers. The presence of other Y haplotypes shared by men over large distances across the Indonesian archipelago, such as Nias, Java/Bali, and Sumba provides more evidence for the Austronesian colonization of the islands. Credit: Authors.

in women as in men. The rest of our chromosomes are autosomes, which are reshuffled in the process of sexual recombination. Consequently we receive about half of our autosomes from each parent. These regions of the genome contain most of our DNA, and they are also informative about ancestry. The more closely related are our parents, the more similar the autosomal DNA that they will transmit to us.

In summary, errors accumulate in the noncoding regions of DNA at predictable rates. This is the basis for molecular clocks and phylogenetic analyses. The branches of a uniparental phylogeny (e.g., patrilines or matrilines) are determined by the number of mutations separating individuals on the tree, using mtDNA for matrilineal descent and NRY variation for patrilineal descent.

The genetic patterns that we see in human communities are the result of kinship practices, which vary between societies. For our purposes the key parameters are the patterns of descent and postmarital residence. Descent can be traced through the mother (matrilineal), the father (patrilineal), or both (bilineal). After marriage, couples can reside with the bride's community (matrilocal), the husband's family (patrilocal), or create a new household (neolocal). In a matrilocal system, women tend to remain in their natal village, where they are joined by their in-marrying husbands. Matrilocal communities usually also trace descent through mothers (matrilineal). In a patrilocal system, the men stay put and the wives marry in; these societies are usually patrilineal.

The implications of matrilocality

The discovery of a sex bias was surprising, but not entirely unexpected. In the 1990s, several studies announced that mitochondrial DNA in Polynesia is predominantly of Asian origin, while Y chromosomes are mostly Papuan.[14] Later studies confirmed this pattern: about 94% of Polynesian mtDNA is of East Asian origin, while about 66% of Polynesian Y chromosomes are Papuan.[15] In 2003, geneticists suggested that this pattern could be explained as an effect of matrilocal residence and matrilineal descent.[16] According to this model, matrilocal Austronesian communities accepted husbands from surrounding Papuan communities. The children of these marriages would inherit their mother's Asian mitochondrial DNA and their father's Papuan Y chromosomes.

However, this hypothesis pertained to Melanesia, where matrilocality is fairly common, not to the islands of Indonesia, where it is rare. And it offered no explanation for why the genetic barrier developed in Wallacea,

---

[14] T. Melton, R. Peterson, A. J. Redd, et al. 1995. Polynesian genetic affinities with Southeast Asian populations as identified by mtDNA analysis. *American Journal of Human Genetics* 57:403–14; A. J. Redd, N. Takesaki, S. T. Sherry, et al. 1995. Evolutionary history of the COII/tRNALys intergenic 9-bp deletion in human mitochondrial DNAs from the Pacific. *Molecular Biology and Evolution* 12:604–15; B. Sykes, A. Leiboff, J. Low-Beer, et al. 1995. The origins of the Polynesians: An interpretation from mitochondrial lineage analysis. *American Journal of Human Genetics* 57:1463–75.

[15] M. Kayser. 2010. The human genetic history of Oceania: Near and Remote views of dispersal. *Current Biology* 20:R194–R201.

[16] The predominance of Asian mtDNA and the high frequency of Melanesian Y chromosomes in Polynesian DNA imply the presence of matrilocal residence and matrilineal descent in Proto-Oceanic society. P. Hage and J. Marck. 2003. Matrilineality and the Melanesian origin of Polynesian Y chromosomes. *Current Anthropology* 44:S121–7.

or for the replacement of Papuan languages. Still, it offered a possible explanation for the sex bias in the DNA: perhaps matrilocality was more common in the distant past. This prompted a question: do matrilocal communities still exist somewhere in Wallacea?

Ethnographic studies mentioned a cluster of matrilocal villages in the mountains of Flores, in a region called Bena. We visited Bena and were able to obtain genetic and linguistic samples, but after sequencing the DNA we found that matrilocality was likely to have been a fairly recent development, no more than a few centuries old. But luckily Bena was not the only candidate. Sixteenth century Portuguese authors described a cluster of matrilocal villages on the island of Timor, in a region called Wehali.[17] A recent ethnography by the late Timorese anthropologist Tom Therik described Wehali as an ancient matrilineal and matrilocal society,[18] organized as a cluster of matrilineal descent groups that engage in marital alliances with each other. Therik also noted that on the borders of Austronesian-speaking Wehali there are villages inhabited by Papuan peoples, who speak Papuan languages. Guided by Therik's ethnography and with the help of his former research assistant, we selected eleven villages in the Wehali region to study, of which eight were matrilocal, while three neighboring villages were patrilocal. In each community we obtained genetic samples from cheek swabs for ~40 men, and double-checked Therik's descriptions of their demographic origins, histories, traditional marriage practices, and census population size. With the help of local public health nurses, we collected demographic and kinship data for each research subject. We learned that the women of Wehali sometimes accept husbands from neighboring Papuan villages, where people speak Papuan languages.

This suggested a possible explanation for the sex bias and the replacement of Papuan languages. Most children in Wehali are born from Austronesian-speaking parents from a Wehali village. But from time to time, Wehali women accept a husband from a neighboring Papuan village. The Papuan villagers speak a Papuan language, and while their genetic ancestry includes both Papuan and Austronesian components as a result of admixture, Y chromosomes are usually Papuan. According to our

[17] T. Therik. *Wehali, the Female Land.* Pandanus Press, Australian National University, 2004.

[18] Both Dutch and Portuguese sources describe Wehali as a matrilineal society, the ritual center of a network of tributary states, from the time of earliest European contact (seventeenth century). Ibid., xv, 4–9. In 2009, Fiona Jordan and colleagues used linguistic reconstruction to argue that matrilocal residence is ancestral in Austronesian societies. F. M. Jordan, R. D. Gray, S. J. Greenhill, and R. Mace. 2009. Matrilocal residence is ancestral in Austronesian societies. *Proceedings of the Royal Society B* 276:1957–64.

informants, these marriages are typically sought by impoverished Papuan men, who by their marriage acquire access to their Wehali wife's property. The children of these marriages inherit their mother's Austronesian mitochondrial DNA and their father's Papuan Y chromosome. Growing up in their mother's village, they speak her Austronesian language.

First model: Sex bias and language replacement

We used what we learned in Wehali to extrapolate a model for the initial Austronesian expansion into Island Southeast Asia. In this model, the Austronesian expansion began with the spread of matrilocal Neolithic societies from Asia into Island Southeast Asia, which was already populated by hunter-gatherers. In eastern Indonesia, these hunter-gatherers were Papuan. As in Wehali today, Austronesian women sometimes accepted husbands from neighboring communities. The children of such marriages inherited their father's Papuan Y chromosome and their mother's mitochondrial DNA. Growing up matrilocally in their mother's village, they spoke her Austronesian language. Our simulation model reflects the following assumptions:

- The initial population of both Austronesian colonists and indigenous hunter-gatherers was small;
- A Neolithic population expansion occurred in each Austronesian village; and
- Small numbers of neighboring Papuan males married into the Austronesian matrilocal houses. The model makes the simplifying assumption that this occurred at a constant rate α.

The predictions of this model are shown in Figure 2.5, in comparison with real data. Even if the migration rate (α) is quite low, there is ample time for a pronounced sex bias to develop. These results may be compared with genetic data from three islands in the region and with data from Polynesia. In the model, the observed genetic patterns will emerge after 50 generations (t) if 2% of marriages are to non-Austronesians. Hence, the model accurately predicts the observed sex bias for all four genetic systems, and also accounts for the replacement of indigenous languages by Austronesian languages. Equations for the model are as follows, where A refers to the autosomes, X to the X chromosome, and Y to the nonrecombining Y chromosome:

$$Y_t = (1 - \alpha)^t \tag{2.1}$$

$$A_t = (1 - \alpha/2)^t \tag{2.2}$$

$$X_t = (1 - \alpha/3)^t \tag{2.3}$$

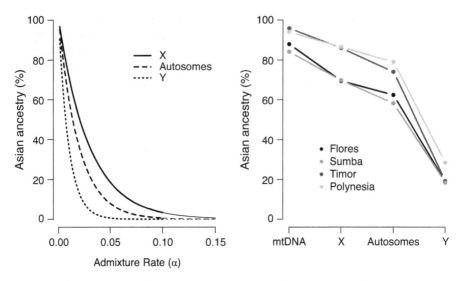

Figure 2.5. Asian ancestry and admixture. (*Left*) Simulation of the average percentage of the genome that is still Asian after 100 generations of Papuan admixture at different admixture rates α. Lines indicate the trajectory of admixture on the X chromosome (solid), autosomes (dashed), and Y chromosome (dotted). (*Right*) Observed percentages of Asian DNA in four genomic systems with decreasing rates of inheritance from women: mitochondrial DNA, the X chromosome, the autosomes, and the Y chromosome. Ancestry percentages are shown for the islands of Flores, Sumba, and Timor in eastern Indonesia, and for comparison, the islands of remote Polynesia. Credit: Authors.

Online Resource: The Austronesian "House" Model

The model of sex bias and language replacement is available to explore in the book's online resources:

https://www.islandsoforder.com/house-societies.html

On Sumba, there have been no matrilocal/matrilineal communities in recent times. However, our model predicts that the original Austronesian colonists in Sumba belonged to a matrilocal/matrilineal house society, and that this form of social organization persisted for many generations. Consistent with this prediction, the concept of matrilineal descent is recognized by all Sumbanese societies, and named matrilineal descent groups still exist in some of them, as noted by van Wouden and later confirmed by Rodney Needham's survey of Sumbanese kinship systems.[19]

Why the barrier in Wallacea?

This simple model provides a concise explanation for the genetic, linguistic, and ethnographic patterns observed in eastern Indonesia and across the Pacific. But it provides no insight into the first question: why did the genetic barrier emerge in Wallacea? A possible explanation begins with the effects of the environment on population movements by Neolithic societies. Earlier studies have proposed that the productivity of rice gardens played an important role in propelling the Austronesians ever further into the islands. It has been suggested that natural climatic variation could underpin the change from rice agriculture to tuber- and palm-based economies, and the decreasing ability of rice horticulture to support the Austronesian expansion into the territory of Papuan peoples.[20] Today rice is uncommon in Wallacea; it is grown in small garden plots in the islands of Wallacea, but does not appear to be a widespread traditional crop. Large-scale irrigated rice cultivation is another matter: it occurs only to the west of Wallacea, on the islands of Java and Bali. While horticulture was practiced by the Austronesians 4,000 years ago, irrigation technology developed much later. Royal inscriptions in Java and Bali indicate that irrigation first appeared on those islands around the sixth century AD.[21] The advent of irrigated rice would have triggered rapid population growth.[22] Still, this explanation is incomplete, because population

[19] R. Needham. *Mamboru: History and Structure in a Domain of Northwestern Sumba.* Oxford University Press, 1987.

[20] Cox, Karafet, Lansing, et al. Autosomal and X-linked single nucleotide polymorphisms, 1589–96.

[21] J. W. Christie. Water and rice in early Java and Bali. In P. Boomgaard, ed, *A World of Water: Rain, Rivers and Seas in Southeast Asian Histories.* KITLV Press, 2007, pp. 23–58.

[22] See J. S. Lansing, A. J. Redd, T. M. Karafet, TM et al. 2004. An Indian trader in ancient Bali? *Antiquity* 78:287–93: "We present the results of a simulation experiment designed to estimate this probability with the following assumptions concerning the population of Bali 2150 years ago. Following Lodewycksz (1597), Reid (1988:13–14; 1999: 199) accepts a population for Bali of 600,000 in 600 AD. A growth rate of 0.12 per cent p.a. from 150 BC is consistent with this estimate. This would produce a population density for the contemporary agricultural regions of Bali (3266 km^2 of sawah and dry lands agriculture) of 13.5 persons/km^2, slightly higher than Kirch's (1984:98) estimate of 12 persons/km^2 for the Hawaiian population at the time of contact. We assumed (1) a population size of 75,000 (with a minimum of 44,000 and a maximum of 220,000) and (2) random mating. We simulated a Wright-Fisher model

growth alone will not explain the sex bias or why the genetic barrier originates east of Bali (where Wallacea begins).

To investigate this conundrum, once again we turned our attention to the effects of kinship practices on the genetic composition of communities. Irrigated rice terraces are permanent valuable assets, and the question of who will inherit them is a matter of great interest to their owners. In Bali, inheritance is traditionally patrilineal. To keep the inheritance of farms within the family, couples without sons in rice-growing villages frequently adopt a nephew, or make their son-in-law their heir. For the same reason (retaining the farms) there is also a strong preference for men to marry a close patrilateral cousin, usually from the same neighborhood or village. In a survey of 252 men in 13 rice-growing Balinese villages, 84% married within their natal village. This preference fell to 34% in two highland Balinese villages, where rice is not grown.[23]

The regions of Bali where irrigated rice is grown today have high population density and very little population mobility. If these characteristics began to appear at the time of the transition from horticulture to irrigated agriculture (circa AD 600), they would have brought population movement into Wallacea to a standstill. This pattern contrasts sharply with marriage practices to the east, in the tribal societies of Wallacea. There, marriages are seldom to the girl next door, as in Bali and Java. Instead they enable individuals and clans to create useful kinship alliances with their neighboring populations. But how long have these contrasting patterns of kinship coexisted?

Second model: Demographic skew

Using genetic samples, it is possible to infer the effects of marriage customs from the demographic history of a community for many generations in the past. Here's how it works:

Every man's genome includes the Y chromosome he inherits from his father and the mitochondrial DNA he inherits from his mother. For simplicity, assume a very small community in which everyone is descended from one of ten grandmothers, and consider the following cases:

1. The grandmothers were sisters; or
2. The grandmothers were unrelated.

---

with maternal inheritance assuming an effective population size of 45 per cent of the total female population size. We assume that the total population of Bali remained constant from 2150 BP until the introduction of irrigated rice around 600 AD, and then rose exponentially to the present population of three million."

[23] J. S. Lansing. *Perfect Order: Recognizing Complexity in Bali.* Princeton University Press, 2006, p. 211.

In case 1, the mitochondrial DNA of our sampled men is identical, because it is inherited from the same great-grandmother. In case 2, it is maximally diverse. In genetics, the concept of *effective population size* refers to the number of ancestors for a given population. In this example, the effective population size on the mother's side for case 1 is 1, and for case 2 it is 10. Effective population size is written N_e and can be calculated for matrilineal ancestors of a community using the mitochondrial DNA, and for patrilineal ancestors using the Y chromosome. These calculations are estimates and are subject to error. But every man has both mitochondrial DNA and a Y chromosome. By comparing their ancestries with those of the other men in his community, it is possible to discover the effects of marriage customs in the past. Patrilocal villages that import wives from elsewhere will have higher diversity of female ancestors, which will show up as greater diversity of mitochondrial DNA. The reverse is true for matrilocal villages, where women remain together and import husbands. But if the community is endogamous, there will be little or no difference in the N_e of male or female ancestors.

Table 2.1 and Figure 2.6 display this comparison—the long-term effects of marriage rules on the genetic composition of villages—across the archipelago from west to east, from the patrilocal and patrilineal villages of Nias, to the endogamous villages of Java and Bali, the patrilocal villages on Sumba, and the matrilocal, matrilineal villages of central Timor. Overall, the Balinese villages are very endogamous. The exceptions that prove the rule are the highland villages in Bali (where rice is not grown), which skew patrilocal. This model is even simpler than the first model: merely a comparison of effective populations of patrilineal and matrilineal ancestors inferred from a sample of men from the village.[24]

Generations of butterfly effects

It is clear that the sharp genetic cline across Wallacea is the result of social processes rather than geographic barriers, because it occurs along a continuous chain of islands that have been populated for thousands of years. We explain this cline by means of two models.

---

[24] Detailed population records are not available for ancient Bali, but for the present discussion it is sufficient to characterize the broad trends. The earliest population estimates for Bali were proposed by Willem Lodewycksz in 1597, about a thousand years after the beginning of wet rice cultivation. Following Lodewycksz, Anthony Reid proposes that Bali had a population of 600,000 in AD 1600. See A. Reid, *Southeast Asia in the Age of Commerce 1450-1680. Volume 1. The Lands below the Winds*. Yale University Press, 1988.

Table 2.1

Differences between effective population sizes (see text) calculated from haplotype data for the Y chromosome and mitochondrial DNA for villages on Sumba and in the Wehali region of central Timor. Note that in the patrilocal communities of Sumba, women have historically had higher effective population sizes than men (positive difference values); in the matrilocal region of Timor, men have had larger population sizes than women (negative difference values).

| Island | Village | Effective Population Size N_e | | |
		mtDNA	Y Chromosome	Difference
Sumba	Anakalang	585	174	411
	Bilur Pangadu	1,421	182	1,239
	Bukambero	786	123	663
	Kodi	554	120	434
	Lamboya	671	111	560
	Loli	345	301	44
	Mahu	1,319	280	1,039
	Mamboro	428	215	213
	Mbatakapidu	643	166	477
	Praibakul	708	198	510
	Rindi	5,274	549	4,725
	Waimangura	404	93	311
	Wanokaka	676	273	404
	Wunga	267	105	162
Timor	Besikama	775	981	−206
	Fatuketi	290	757	−467
	Kakaniuk	147	450	−303
	Kamanasa	1,178	1,744	−566
	Kateri	366	397	−31
	Kletek (all villages combined)	390	1,416	−1,026
	Laran	863	1,347	−484
	Tialai	274	442	−167
	Umaklaran	219	859	−641
	Umanen Lawalu	272	1,416	−1,144

In Bali and Java, located to the west of Wallacea, a population explosion triggered by the spread of irrigated rice cultivation erased most genetic traces of earlier Papuan hunter-gatherer populations. As sedentary, highly endogamous farming communities grew up along the rivers and irrigation canals of Bali, few people left home and gene flow to the east came to a standstill.

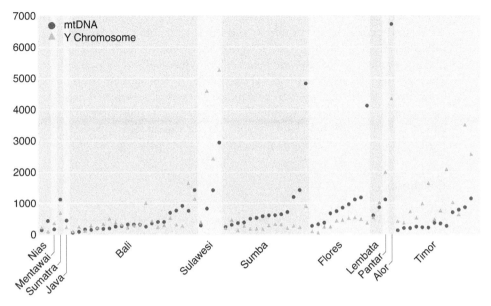

Figure 2.6. Differences between effective population sizes across the entire archipelago, from west to east. In patrilocal villages, women marry in and their effective population size is higher. The reverse is true for matrilocal communities, where husbands migrate in. These differences can be calculated from the diversity of mtDNA and Y chromosome haplotypes (see text). They reflect the long term consequences of kinship systems as well as other historical processes, such as reproductive skew (the subject of chapter 3). Credit: Guy Jacobs.

Eastwards from Bali into Wallacea and beyond, it is clear from the genetic and archaeological evidence that the islands were already populated when the Austronesians began their colonizing voyages. Our model provides an explanation for both the progressive sex bias in Asian DNA and the replacement of earlier indigenous languages with Austronesian languages. It is based on three assumptions: the initial population of both indigenous peoples and Austronesian colonists was small, the Austronesian colonies grew quickly thanks to their skill as farmers, and Austronesian women occasionally married men from surrounding non-Austronesian villages. The children of such marriages would have spoken their mother's Austronesian language, while the total population of Austronesian speakers steadily increased. This model suggests that over a time scale of tens of generations, a seemingly trivial shift in marriage preferences can produce a seismic change in language, culture, and demography. The data we have reviewed here provide strong evidence that at a first approximation, this transformative potential was realized in the wake of the Austronesian colonization of Island Southeast Asia.

The idea that such a subtle process could have such profound consequences seems counter-intuitive, and leads us to offer a final remark. Edward Lorenz's "butterfly effect" quickly became the canonical example of the emergence of chaotic behavior in dynamical systems caused by their sensitive dependence on initial conditions.[25] The mathematical basis of the butterfly effect is the Lorenz equations: three first-order differential equations in which the iteration of tiny variation in initial parameterization quickly produces chaotic dynamics. In our model, however, the iteration of small α over many generations created new regimes of order.

Conclusion

The colonization of the Pacific was the last great human expansion into pristine environments. Our goal in this chapter was to sketch the outlines of that history, and set the stage for the chapters to follow. We also wished to make two points about models. The first goal was to show that a very simple model of seemingly complex and unrelated phenomena—genetic and language variation over vast distances—can have significant explanatory power. Until a century ago, the most widespread language family on Earth was Austronesian. Combining genetic and linguistic data, we discovered that centuries of interethnic love affairs played a key role in its expansion. Our second goal was to show how genetic data can be used to recover detailed information about progressive historical changes in phenomena like kinship and marriage.

We are not the first anthropologists to take an interest in sex in the Pacific.[26] In the chapters to follow, we redirect this interest from the entire Indo-Pacific region to island communities. In particular, we will follow up the clue that subtle changes in the relations between the sexes can be a powerful engine of change. Sex is typically entwined with power, as Baudelaire observed, and a scarcity or predominance of ancestral lines in communities offers clues to these interactions. To recover these histories, we need to reconstruct the origins of patrilineal and matrilineal ancestors at the community or village scale. The simplest calculation is the one we described above: a comparison of the effective population size of both sets of ancestors, maternal and paternal. Figure 2.6 shows this comparison for villages on many islands across the Malay archipelago. A glance at this figure shows not one pattern, but many. The question is, why?

---

[25] R. Hilborn. 2003. Sea gulls, butterflies, and grasshoppers: A brief history of the butterfly effect in nonlinear dynamics. *American Journal of Physics* 72:425–7.

[26] See, for example, Bronisław Malinowski, *The Sexual Lives of Savages* (1929), *Argonauts of the Western Pacific* (1922); Margaret Mead, *Coming of Age in Samoa* (1928).

In light of the evidence we have just reviewed about the role of women in the colonization of the Pacific and the rapidity of Austronesian colonization, what do we see at the village and island level? Is there evidence for competition for mates? Changing relations between the sexes? The existence of dominant classes? All of these processes will leave traces that can be read from the genetic record. In the next chapter, we look for evidence of selection, dominance, and competition. In the following chapter, we bring language back into the story.

The genetic evidence is clear: something unusual happened in Bali. But genetics alone does not hold the answer. Three chapters on, we seek an explanation by introducing models of a different kind of dynamical process, one that does not depend on kinship ties, and has the power to weaken or suppress dominance and competition.

~~~~~~

CHAPTER 3

Dominance, Selection, and Neutrality

Territory, status and love
sing all the birds
are what matters.

—*W. H. Auden*

Introduction

The Austronesian seascape encompasses enormous cultural variation, delightfully evoked by anthropologist James J. Fox:

A diversity of island environments has called forth adaptations that have also spawned great social variety: coastal sago palm exploiters such as the Waropen; elusive jungle nomads like the Kubu of Sumatra; tiny fishing populations on islands in the Pacific; riverine peoples such as the Dayak, Kenyah or Kayan of Borneo; the maize-cultivating mountain populations like the Atoni Meto of Timor, for whom a view of the sea was once considered distressing; dryland palm tappers such as the Rotinese and Savunese in eastern Indonesia; cattle herding peoples such as the Bara of south central Madagascar; yam, taro and sweet-potato gardeners of the Melanesian islands, some of whom, like those on Goodenough Island or the Trobriands, flaunt their harvests in feasting for recognition; expansive swidden-rice cultivators like the Iban of Borneo; or settled rice farmers like the Ifugao of Luzon with centuries of collective investment in elaborate terraces.[1]

Can such enormous diversity be explained? In contemporary social science, only one theory is sufficiently broad and ambitious to be a candidate: Darwinian selection. As we saw in the last chapter, comparisons of genetic data at the scale of communities and islands—rather than individuals—make it possible to detect changes in social organization over very long time scales. In this chapter we use data from many of the same communities to pursue a theoretical question: the role of Darwinian selection in social life. Simply put, is Darwin's theory up to the job? The

---

[1] P. Bellwood, J. J. Fox, and D. Tryon. *The Austronesians: Historical and Comparative Perspectives*. Research School of Pacific Studies, Australian National University, 1995, pp. 214–5.

time scale shortens to tens of generations, and the spatial scale shrinks from the entire region to individual communities, each of which is treated as a separate case.

Along with its outstanding pedigree, Darwinism has impressive empirical support. Many studies have argued that reproductive skew is commonly biased toward dominant or high-ranking men in human communities: "In more than one hundred well studied societies, clear formal reproductive rewards for men are associated with status: high-ranking men have the right to more wives."[2] Demographic statistics collected over short time scales support these claims.[3] However, although variation in male fitness is known to occur, an important unanswered question is whether such differences are heritable and persist long enough to have evolutionary consequences at the population level. If not, selection can be ruled out as the cause. Our genetic data from the islands offered a chance to analyze changes in the strength and persistence of reproductive skew in 41 communities, scattered across half a dozen islands.

We begin the chapter with this case study. Unsurprisingly, we find that the role of Darwinian selection in social life is a question for which genetic data are particularly well suited. But while the methods we use here originated in molecular biology, they can also be applied to processes of change in which genetics plays no part. For example, archaeologists are often interested in understanding changes in the distribution of artifacts like tools or ceramics. Similarly, ecologists may wish to know whether differences in the prevalence of species in a region are due to selection or simply to chance. And sociologists might wonder whether generational changes in phenomena like the popularity of children's names are under active selection. In the second part of the chapter, we will examine all three of these cases, not merely as illustrations of the basic concepts, but as a way to show how these models can be adapted to diverse empirical questions. While the conceptual framework for distinguishing selection from drift is simple and intuitive, estimating the strength and power of these processes in specific cases is a little more complicated. Hence we have two goals for this chapter: first, to assess the capacity of Darwinian selection to explain the evolution of social organization in Indonesian communities, and second, to explore how selection can be detected in other kinds of evolving populations.

---

[2] A. L. Clarke and B. S. Low. 2001. Testing evolutionary hypotheses with demographic data. *Population and Development Review* 27:633–60.

[3] B. Winterhalder and E. Smith. 2000. Analyzing adaptive strategies: Human behavioral ecology at twenty-five. *Evolutionary Anthropology* 9:51–72.

Selection for dominance?

Evolutionary social scientists analyze the fitness consequences of behavior, for which the currency of fitness is reproductive success (i.e., a selective advantage results in more descendants). Although the inspiration for this question is Darwinian evolutionary theory, biology is not conceptually embedded in the methods. But as we will see, genetic markers can be used to discover whether selection has occurred in real populations by examining the distribution of genetic diversity. Importantly, this technique is not based on identifying the selective advantage of specific genes. Instead, neutral markers from noncoding regions of the genome are used to identify branching patterns in family trees (uniparental lineages), which can in turn be used to discover whether there is a tendency for some lineages to become more numerous by acquiring a selective advantage.

Figure 3.1 shows how this is done. If selection is not present (i.e., the system is neutral), every individual has an equal chance of producing offspring (Figure 3.1a). But if some individuals obtain a heritable reproductive advantage, their descendants will become disproportionately abundant in the population (Figure 3.1b). A third possibility exists: some individuals in each generation might attain higher fitness (i.e., have more children) but not pass this characteristic on to their own children. This produces a "Red Queen" dynamic, in which a selective advantage occurs but seldom persists within families (Figure 3.1c). The Red Queen forestalls evolutionary change by preventing any descent group from gaining a lasting advantage. As she explained to Alice, sometimes "it takes all the running you can do to keep in the same place."[4]

To apply this conceptual model to an empirical question, we need to reconstruct the patrilineal descent of a representative sample of male villagers. The Y chromosome makes this easy because it is passed directly from fathers to sons. As in all genetic systems, random mutations occur

---

[4] Originally proposed by Leigh van Valen and commonly conceptualized in the context of host-parasite arms races, the Red Queen hypothesis states that continuing adaptation is needed just to maintain fitness. That is, in the context of an antagonistic coevolutionary arms race, evolution is occurring even though mean fitness remains nearly stationary. At its most simple, a Red Queen dynamic is simply the interaction between two (or more) competing evolutionary units. Here, we merely extend this definition to include different "families" (i.e., independent paternal lineages). The significance of the Red Queen model is that it defines a particular form of evolutionary change that is measurably distinguishable from related models, such as simple directional selection or neutral drift. The nonheritable variation in male reproductive success described here results in reduced effective population size compared to neutral equilibrium, but not as much as would occur as a result of simple dominance. L. van Valen. 1973. A new evolutionary law. *Evolutionary Theory* 1:1–30.

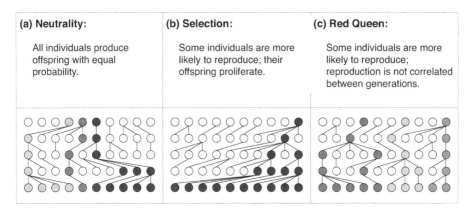

(a) Neutrality:	(b) Selection:	(c) Red Queen:
All individuals produce offspring with equal probability.	Some individuals are more likely to reproduce; their offspring proliferate.	Some individuals are more likely to reproduce; reproduction is not correlated between generations.

Figure 3.1. Noncoding genetic markers can be used to track lines of descent; here, each shade represents a single haplotype. Populations are (a) at neutral equilibrium; (b) undergoing selection; (c) experiencing Red Queen dynamics in which dominance fluctuates and high fecundity is not inherited. Credit: Brian Hallmark.

in the noncoding regions over time. These mutations have no functional or behavioral significance, are not under selection, and so merely provide a genetic record to trace the relationships of men over time. Men who are closely related to each other share similar patterns at these noncoding sites. The most closely related men show no variation at all, and are defined as sharing an identical Y chromosome haplotype, indicating that they are very close patrilineal relatives.[5]

Haplotypes are routinely used for research in population genetics. However, most human genetic studies sample haplotype variation from a broad geographical catchment, often visitors at a regional medical clinic, and thus include many communities. Because our goal was to analyze changes in haplotype frequency at the community scale, we collected representative samples from well-defined communities that have existed for generations. This change in scale, from the fecundity of individuals over a generation or two to the distribution of haplotypes within communities, made it possible to distinguish selection from random drift or Red Queen competition. For example, if Lineage A has high social rank, which

[5] To construct haplotypes, we genotyped 12 microsatellites—regions of the DNA that vary by length—and a battery of single nucleotide polymorphisms—single base changes—on the Y chromosomes of 1,269 men from 41 Indonesian communities. For details, see the supplementary materials section of J. S. Lansing, J. C. Watkins, B. Hallmark, et al. 2008. Male dominance rarely skews the frequency distribution of Y chromosome haplotypes in human populations. *Proceedings of the National Academy of Sciences USA* 105:11645–50.

translates into more sons, who in turn pass on their greater fecundity (fitness) to their own descendants, then over time Lineage A will become more abundant in the community. Communities experiencing reproductive skew among patrilines will develop haplotype frequency distributions that become increasingly unlikely under the neutral model. This is easily detected, because there is only one neutral frequency distribution for any given data set and population model. The neutral distribution depends solely on the total population size and the rate at which new variants (in this case, haplotypes) appear. If selection is present, it will cause departures from the neutral distribution that are readily detectable if the sample size is adequate.

However, departures from neutrality are not solely due to selection; they can also arise from demographic processes. The effective population size, N_e, is proportional to the number of haplotypes in a given population. To find out whether the haplotype distribution in a community is at neutral equilibrium, it is enough to know this number and the mutation rate (the rate at which new Y chromosome haplotypes appear). Figure 3.2 plots simulations of neutral equilibrium for different-sized communities. Like the marbles in the statistician's bag, as time goes forward, chance will cause some haplotypes to become more numerous while others dwindle and disappear. Neutral equilibrium occurs when these rates balance. New haplotypes appear whenever a mutation appears in an existing haplotype, or when newcomers marry into the community. If no new haplotypes appear, eventually only one will remain. The larger the population, the longer it takes for this process to reach equilibrium.

Interestingly, any deviations from neutrality in the population persist for a long time. Figure 3.2 shows the return to neutrality from two extreme situations: all haplotypes initially identical (lower curves in each panel) and all haplotypes initially different (upper curves in each panel). For an average-sized Indonesian village (say, effective size $N_e = 400$), non-neutral diversity persists for ~20 generations or ~600 years following a selective event, such as an instance of male dominance.[6]

What gives the neutral test its power is the fact that the distribution of haplotypes at neutral equilibrium forms a characteristic pattern. The chance appearance and disappearance of haplotypes causes their frequency to drift in predictable ways: a few will become abundant, while more will become rare or go extinct. Given two assumptions, that the population size is constant and that each person has an equal chance to

---

[6] Assuming a male generation interval of 30 years; see J. N. Fenner. 2005. Cross-cultural estimation of the human generation interval for use in genetics-based population divergence studies. *American Journal of Physical Anthropology* 128:415–23.

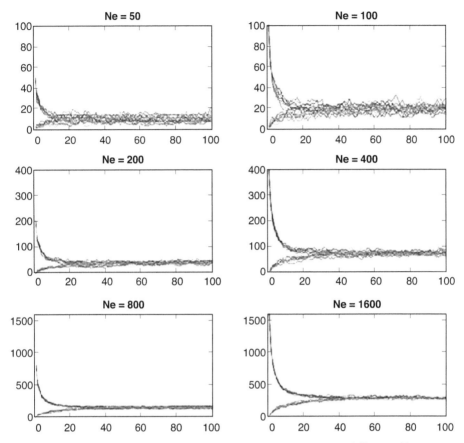

Figure 3.2. Persistence times of non-neutral genetic diversity at different effective population sizes, N_e. Time in generations is shown on the x-axis, effective population size on the y-axis. Initial conditions are extreme cases: either all haplotypes unique (upper curves) or all haplotypes identical (lower curves). The mutation rate m in these simulations is 0.0208 per generation, based on calculations for the Indonesian samples described in the text; the male generation interval is 30 years. For an average-sized population (N_e = 400), non-neutral diversity can be observed for ~20 generations (~600 years) following a selective event. The finding that 88% of our Indonesian populations exhibit neutral diversity is therefore doubly interesting. Not only are most of these populations neutral today, but there is little evidence for selection in the past. Credit: Brian Hallmark.

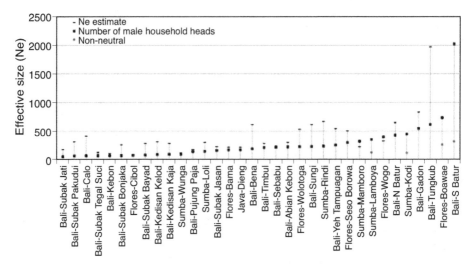

Figure 3.3. Estimates of effective male population size, N_e, for sampled communities, together with the number of male household heads. Data are sorted by census size, and six sites for which census data were not available have been excluded. Dashes above squares indicate N_e estimates that are larger than census sizes; dashes below squares indicate a reduction in N_e. Gray circles indicate sites that tested as non-neutral. Of these five communities, four showed a reduction in N_e and one site lacked census data. The data point for South Batur, indicating an effective male population size of ~2,000, is correct, for reasons explained in the text. Credit: Authors.

reproduce, Warren Ewens showed that the resulting neutral frequency distribution is predicted by the Ewens sampling formula.[7] How closely a real empirical distribution approximates the Ewens distribution can be tested using Slatkin's exact test, as developed by Montgomery Slatkin (Figure 3.3) (see boxed inset).[8]

---

[7] We used the Ewens–Watterson equation to calculate the neutral distribution of patrilines for each village. This equation is powerful because it makes precise predictions based on simple demographic assumptions: constant population size and equal probability that any individual will reproduce. For this reason, versions of this equation are widely used in both genetics and ecology. W. Ewens. 1972. The sampling theory of selectively neutral alleles. *Theoretical Population Biology* 3:87–112.

[8] The Slatkin Exact Test evaluates the empirical frequency distribution against all possible configurations for a given *n* and *k* under the Ewens sampling distribution, and is a more general test that makes fewer assumptions about which aspects of distribution shape might indicate selection: M. Slatkin. 1994. An exact test for neutrality based on the Ewens sampling distribution. *Genetic Research* 64:71–4; M. Slatkin. 1996. A correction to the exact test based on the Ewens sampling distribution. *Genetic Research*

Ewens Sampling Formula

The Ewens sampling formula, introduced by Warren Ewens in 1972, describes the frequency and distribution of *types* within a sample. In genetics, these types are typically alleles (*haplotypes*), but in other settings, they could be different classes of axe head or different species of tree.

As we saw earlier, when neutrality holds, the diversity of a sample is given by the key parameter θ ($= 2N\mu$). A close relationship exists between the diversity of a sample, as measured by θ, and the number of different types that sample contains. Ewens's contribution was to state this relationship mathematically, deriving the maximum likelihood estimator of the number of types k:

$$E(k) \cong \sum_{i=0}^{n-1} \frac{\theta}{\theta + i} \tag{3.1}$$

We can use this number of types, which is typically easy enough to just count, to estimate the value of θ. One way to do this is to screen a range of θ values, calculate the expected value of k for each θ, and choose the θ value that returns a k that best matches the observed count.

Ewens also derived an equation that describes the distribution of types within a sample (usually called the *haplotype frequency spectrum* in genetics):

$$E(k; x_1, x_2) = \theta \int_{x_1}^{x_2} x^{-1}(1 - x)^{\theta-1} \, dx, \text{ where } N^{-1} \leq x_1 \leq x_2 \leq 1 \tag{3.2}$$

This distribution only holds under neutrality, but because of this, it provides the basis for a formal test of selection. If our data set has a distribution that differs from this neutral expectation, then we can make a case that selection may have acted on the data set.

68: 259–60. Slatkin has helpfully archived online his source code for a Monte Carlo simulation program that conducts both tests (Ewens–Watterson and Slatkin's Exact). The outputs are tail probabilities, which indicate the position of an empirical set of data in a probability distribution of possible sets derived from the Ewens sampling formula: https://ib.berkeley.edu/labs/slatkin/monty/Ewens_exact.program.

The key is to ask how likely some distribution of types that we observe would be under neutrality, compared to all other possible distributions given the same sample size n and number of types k. To calculate this, we need to know the likelihood of observing a given distribution, which Ewens also determined:

$$Pr(n_1, \ldots, n_k | k, n, neutrality) = \frac{n!}{k! \, l_k n_1 \cdot n_2 \cdots n_k} \quad (3.3)$$

where n is the sample size and $l_{k,n}$ are special values used in combinatorics, called *unsigned Stirling numbers of the first kind*.

This equation forms the basis of Montgomery Slatkin's exact test, which works by calculating the probability of each possible distribution, and bins them according to whether the probability is greater or smaller than that of the observed distribution. In this way, it is possible to make a statistical statement about whether the distribution of types seen in a sample likely results from selection or not—regardless of whether they are alleles, axes, or trees.

W. J. Ewens. 1972. The sampling theory of selectively neutral alleles. *Theoretical Population Biology* 3:87–112.

Online Resource: Neutrality

The model of selection is available to explore in the book's online resources:

https://www.islandsoforder.com/neutrality.html

The meek shall inherit ...

When we began this study, we wondered what might cause selection for male dominance to vary. Perhaps competition between patrilines might be more intense in patrilocal communities, where grown sons remain in their father's village, compared to matrilocal communities, where grown sons join their wife's household? Or mobile hunter-gatherers whose survival depends on cooperation might be less prone than tribal farmers to compete for dominance? And yet... perhaps hunter-gatherers might nonetheless compete for wives? "It is easy to invent a selectionist explanation for almost any specific observation," observed an exasperated Motoo Kimura. "Proving it is another story. Such facile explanatory excesses can be avoided by being more quantitative."[9]

Heeding Kimura's advice, we used Y chromosome haplotypes to test for selection in 41 communities on nine islands, including nomadic hunter-gatherers on Borneo, matrilocal farmers on Flores, patrilocal Balinese rice farmers, neolocal Javanese rice farmers, and patrilocal clans on Sumba, Nias, and Flores. To our surprise, we found that all but five of these communities are at neutral equilibrium with respect to their distribution of Y chromosome haplotypes.[10] Three of the five non-neutral villages are consistent with positive selection; they potentially experienced male dominance. However, the other two instead appear to be skewed by recent demographic processes; rather than haplotypes being too similar, they are actually too diverse. The remaining 36 villages in the study turned out to be at neutral equilibrium. This does not imply that social dominance or competition between males does not occur in these communities, but it does mean that no heritable traits or behaviors that are passed paternally, be they biological or cultural, were under selection strong enough to have detectable evolutionary consequences.

The three villages that showed evidence for selection are on Sumba, a remote eastern island where the proportion of the population that practices a traditional tribal religion is the highest in Indonesia. Sumbanese villages are patrilocal and descent is traced through the patriline. Marriages are polygamous, and competition for status and resources among patrilineal clans is endemic.[11] Given these circumstances, it is perhaps remarkable that the other five Sumbanese communities we sampled failed to show evidence of heritable reproductive skew.

We were also surprised to find that all but one Balinese village was at neutral equilibrium. Like Sumba, Balinese villages are patrilocal, and there is also plenty of competition among patrilineal descent groups. But

[9] M. Kimura. *The Neutral Theory of Molecular Evolution.* Cambridge University Press, 1983, p. xiv.

[10] For full methods, see Lansing, Watkins, Hallmark, et al., Male dominance, 11645–50.

[11] G. L. Forth. *Rindi.* Martinus Nijhoff, 1981; J. Hoskins. *The Play of Time: Kodi Perspectives on Calendars, History and Exchange.* University of California Press, 1993; J. S. Lansing, M. P. Cox, S. S. Downey, et al. 2007. Coevolution of languages and genes on the island of Sumba, eastern Indonesia. *Proceedings of the National Academy of Sciences USA* 104:16022–6.

the Y chromosome haplotype frequencies in all of the Balinese villages were at neutral equilibrium, with the exception of South Batur. This village was once part of a larger village called Batur. In 1948, after a period of rivalry between factions, Batur split into two communities, North and South, and many families found it necessary to relocate.[12] This recent demographic history, rather than reproductive skew, explains the unusual abundance of Y chromosome haplotypes in South Batur.

The remaining non-neutral community is Boawae, a patrilineal and patrilocal community located in central Flores. It was formerly the site of a minor princedom that became an administrative center during the Dutch colonial era and now serves as a district capital. Our sample—on looking closer—included a large proportion of civil servants born elsewhere, which accounts for the extra haplotypes in the population. Apart from these five villages, none of the sampled populations showed evidence of a departure from neutral stochastic equilibrium with respect to male lineages.

Summing up, if reproductive skew inherited between generations were a pervasive and ongoing process, we would expect to observe frequent rejections of the Ewens distribution of neutral variation. We do not observe this in the vast majority of our Indonesian communities. Only in Sumba did we find a few villages in which haplotype distributions are consistent with selection for dominance. We conclude that male reproductive skew is at best weak in the other communities, despite their varied locations, subsistence strategies, and kinship practices.

Neutral tests and the neutral theory: From genetics to ecology

These results are enough to raise a doubt about the popular Darwinian explanation for male dominance, although they are hardly conclusive. We followed up with more statistical tests and simulations, to assess the possible consequences of varying the demographic scenarios and/or strength of selection.[13] It turned out that even short episodes of weak selection would leave a detectable signature in the distribution of haplotype frequencies. Encouragingly, across a wide range of demographic scenarios, the test was able to discriminate clearly between

[12] J. S. Lansing. *Perfect Order: Recognizing Complexity in Bali.* Princeton University Press, 2006, pp. 179–85.

[13] Lansing, Watkins, Hallmark, et al. Male dominance, 11645–50.

regimes of neutrality and non-neutrality caused by either selection or population movements.

Online Resource: Selection

The model of neutrality is available to explore in the book's online resources:

https://www.islandsoforder.com/selection.html

Still, how far should we generalize from such results? In genetics, Kimura used his statistical test for neutrality to support a *neutral theory*, which holds that most evolutionary change is the result of drift rather than selection. But this conclusion is not baked into the statistical tests for neutrality. Kimura's first paper on neutrality was published in 1968, prompted by the discovery that the genetic diversity revealed by early forms of gel electrophoresis appeared to be much too high to be explainable by natural selection.[14] Kimura's neutral theory proposed that selectively neutral variants (alleles) arise by mutation and then fluctuate in abundance randomly.

Kimura consistently emphasized that his neutral theory applied to molecular diversity, not phenotypes. But in 1976, ecologist Hal Caswell suggested that Kimura's statistical model for neutral evolution could be adapted to analyze the role of selection in ecological communities. Kimura had asked: what would molecular diversity look like if it were the result of drift rather than selection? Caswell proposed to use Kimura's statistical model to try out the same question for species diversity. What if species originate at random and their abundances simply fluctuate randomly over time?[15]

It was not long before these tests were undertaken. But before going on to discuss the role of neutral tests in ecology, three points are worth noting. First, Kimura's model is easily adapted, and was indeed the obvious choice, because it is the simplest possible model for pure chance in an evolving population. Second, in both genetics and ecology, no one doubted that selection played some part in evolutionary change. The point of the neutral tests was to quantify how much. Yet in both

---

[14] M. Kimura. 1968. Evolutionary rate at the molecular level. *Nature* 217: 624–6.

[15] H. Caswell. 1976. Community structure: A neutral model analysis. *Ecological Monographs* 46:327–54.

fields, the neutral theory quickly became very controversial. Our third point is to suggest a reason for that controversy. In both genetics and ecology, the neutral theory requires the same shift in focus, from the selective advantages of specific mutations to the macroscopic properties of communities. Mutations are studied from a traditional Darwinian perspective on a case-by-case basis, but neutrality is a statistical property. Ideal cases do not exist, and neither do neutral ecological communities or genotypes. To appreciate the heuristic value of these tests therefore required a shift in perspective, one much closer to physics than traditional biology.

By now that shift has been accomplished in genetics, which is increasingly a statistical science. Today Kimura's equations are seldom invoked by geneticists to argue the merits of the neutral theory. Instead they are routinely used to identify and measure the strength of selection. In ecology, interest in neutrality picked up in the 1990s when neutral tests began to be applied to large data sets. In response to a seemingly unresolvable plethora of explanations of how so many kinds of trees can coexist in communities, especially in tropical settings, ecologist Stephen Hubbell formulated an ecological version of the neutral theory to explain biodiversity patterns without invoking selection on species differences. In Hubbell's neutral theory, each tree's prospects for reproduction and death are assumed to be independent of its species or that of its neighbors. This assumption is simple enough to allow neutral theory to unify diverse aspects of ecology and biogeography, such as species abundance distributions, changes in species composition over space and time, and the impacts of habitat fragmentation in a "unified neutral theory of biodiversity and biogeography." The analysis is carried out in guilds, like trees or corals, within well-defined communities.[16] Analytically, this method is essentially identical to the approach we used to study Y chromosome haplotype distributions in the Indonesian villages. The two neutral theories, for population genetics and ecology,[17] are compared in Table 3.1.

[16] S. P. Hubbell. 1997. A unified theory of biogeography and relative species abundance and its application to tropical rain forests and coral reefs. *Coral Reefs* 16:S9–21; S. P. Hubbell. *The Unified Neutral Theory of Biodiversity and Biogeography*. Princeton University Press, 2001.

[17] D. Alonso, R. S. Etienne, and A. J. McKane. 2006. The merits of neutral theory. *Trends in Ecology and Evolution*. 21:451–7.

Table 3.1
The extension of Kimura's neutral theory from genetics to ecology.

Property	Community ecology	Population genetics
System (size)	Metacommunity (J)	Population (N)
Subsystem	Local community	Deme
Neutral system unit	Individual organism	Individual genetic locus
Diversity unit	Species	Allele
Stochastic process	Ecological drift	Genetic drift
Dispersal	Immigration m	Migration m
Innovation	New species v	New mutations μ
Fundamental diversity number	$\theta \approx 2J_M v$	$\theta \approx 2N\mu$
Fundamental dispersal number	$I \approx 2J_L m$	$\theta \approx 2Nm$
Relative abundance distribution $\Phi(x)$	$\theta x^{-1}(1-x)^{\theta-1}$	$\theta x^{-1}(1-x)^{\theta-1}$
Time to common ancestor (small θ approximation)	$-J_M x(1-x)^{-1}\log(x)$	$-Nx(1-x)^{-1}\log(x)$

Transience and time scales: Baby names

The successful adaptation of tests for neutrality from genetics to ecology points to an important conclusion: any process that can be modeled using replicators will leave a signature in the population that can be tested for neutrality. But both geneticists and ecologists have the luxury of thinking in terms of vast evolutionary timescales. In the past decade, archaeologists and other social scientists have begun to experiment with neutral models for cultural phenomena, which play out over much shorter time scales and so might never reach equilibrium. In the Indonesian case we have just considered, on a time scale of centuries, equilibrium was the rule and transient dynamics the exception. If we shorten the time horizon even more, is it still possible to distinguish selection from neutrality?

In 2002, the US Social Security Administration published the thousand most common baby names in each decade of the twentieth century, based on a sample of 5% of all social security cards issued to Americans. From the perspective of the parents (including us), choosing a name for a child is a matter of lengthy deliberation—careful selection, not chance. These choices might be expected to reflect enduring trends or fashions; in each generation parents might tend to select the names of culturally dominant or prestigious individuals. The choice of names might also be influenced by the religion or ethnicity of the parents. Any number of selectionist hypotheses for the prevalence of specific names could be imagined for this data set. The alternative, neutral hypothesis would predict a distribution of names that appears no different from that produced solely by chance.

Matthew Hahn and Alexander Bentley[18] used Kimura and Crow's method to test the neutral hypothesis (see boxed inset). The data consist of the cumulative frequency distribution of the 500 most popular names in each decade for boys and girls. To compare the empirical distributions of names with the predictions of the neutral model, they created a simulation in which the invention of a new baby name is analogous to mutation. Each time step in the simulation represents a set of N newly born babies, each of whom is named by copying the name of a randomly chosen baby from the previous time step. In addition, in each time step a small fraction μ of the N individuals receive a unique name.[19] Both the empirical data and the simulation are consistent with neutrality, and the model accurately predicts the empirical distribution of names at each decadal time step (Figure 3.4).

Kimura and Crow and Neutrality

In 1964, Motoo Kimura and James Crow made a major breakthrough in our understanding of neutral systems. They derived an equation that shows how many neutral variants—such as genetic alleles—with frequency x should exist in a population, given it has size N and mutation rate μ:

$$\theta x^{-1}(1 - x)^{\theta-1}, \text{ where } \theta = 2N\mu \qquad (3.4)$$

Crucially, this equation does not hold just for genes; it applies equally to any neutral distribution of objects—tree species in the Amazon, pot types in an archaeological assemblage, and the names of American babies.

While Kimura and Crow were scribbling on their blackboard, the statistical ecologist Carrington Williams was coincidentally counting

---

[18] M. W. Hahn and R. A. Bentley. 2003. Drift as a mechanism for cultural change: An example from baby names. *Proceedings of the Royal Society B* 270: S120–3.

[19] This simple model is equivalent to the infinite allele model of population genetics for a single-locus, multiple neutral-allele system; M. Kimura and J. F. Crow. 1964. The number of alleles that can be maintained in a finite population. *Genetics* 49:725–38; J. F. Crow and M. Kimura. *An Introduction to Population Genetics Theory*. Harper Row, 1970.

beetle species in the River Thames.[20] When the distribution of beetle species is graphed—here as a Preston plot, named after the English-American ecologist Frank Preston—it shows a characteristic pattern.

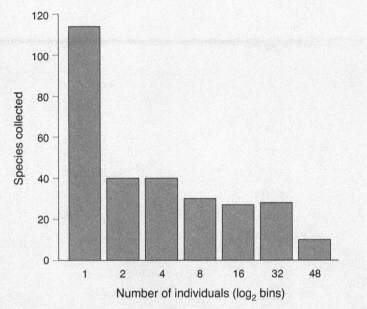

For most species, only a single individual was found. Beetles were more frequent for a few other species, but for only a very small number of species were beetles common. This distribution is called a *power law*.

Power laws are found time and time again in nature, and are highly characteristic of neutral data sets—systems that have not been acted on by selection. Unbeknown to each other, Kimura, Crow, and Carrington were discovering the same fundamental rule of nature. Fifty years later, we observed this neutral pattern in the Y chromosome diversity of Indonesian men, despite generations of struggles for dominance.

M. Kimura and J. F. Crow. 1964. The number of alleles that can be maintained in a finite population. *Genetics* 49:725–38.

C. B. Williams. *Patterns in the Balance of Nature and Related Problems in Quantitative Ecology.* Academic Press, 1964.

[20] Given the state of the Thames in the 1960s, we are surprised he found any at all.

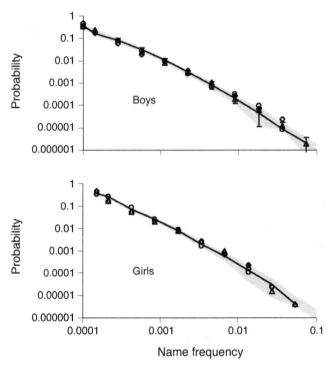

Figure 3.4. Power law distributions of baby names. Frequencies of the most common 1,000 male (top) and female (bottom) baby names in the United States, for four representative decades during the twentieth century (1900–1909, 1940–1949, 1950–1959, and 1980–1989). The x-axis represents the frequency of a name in the total sample of individuals; the y-axis represents the probability that a certain name falls within the bin at that frequency (proportional to the fraction of names in the top 1,000). As is common for such log-log plots, the bin sizes increase in powers of 2 (e.g., 0.0001–0.0002, 0.0002–0.0004, 0.0004–0.0008, ...). Data are plotted at the middle of each bin and probabilities are normalized for the increasing bin sizes. Also shown are the mean (solid line) and 95% confidence intervals (gray ribbon) resulting from 20 runs of the neutral-trait model with $\theta = 2Nm = 4$. A regression between log(average model value) and log(average data value) yields $r^2 = 0.993$ for male names and $r^2 = 0.994$ for female names. Credit: Modified from M. W. Hahn and R. A. Bentley. 2003. Drift as a mechanism for cultural change: An example from baby names. *Proceedings of the Royal Society* B 270:S120–3.

While both boy's names and girl's names were at neutral equilibrium at the decadal time scale, one clear difference emerged: the rate of the introduction of new names was always higher among girls (2.3 names per 10,000 births) than boys (1.6 names per 10,000 births). Hahn and Bentley suggest that this difference may be the result of naming customs in a predominantly patriarchal society, and also the fact that only about 6% of all the names in Judeo-Christian scriptures are female. Overall, they conclude that "the distribution of baby names in the United States and their turnover in the twentieth century can be explained by a completely value-neutral process: no preference, fitness or selection is needed, only proportional sampling."[21]

However, the absence of selection does not imply the absence of change. Two follow-up studies provide clear illustrations of why this is important. Recall that the neutral model requires two parameters, population size N and the mutation rate μ (here interpreted as the appearance of new names). In Hahn and Bentley's study, N was the entire population. Paolo Barucca and his colleagues broke the data down by states to check for patterns at that scale. To do so they cross-correlated the data to find out whether names in one state are correlated with names in another state.[22] They found evidence for a neutral diffusion process, in which neighboring states tend to share similarities in preferences for names. On the decadal time scale, the strength of these correlations—and resulting patterns of diffusion—gradually shifts. The second study found a rapid acceleration in μ around the turn of the twentieth century, with estimates of μ more than doubling for girls and tripling for boys by the end of the first decade of the twenty-first century. At the scale of the states, they found that in 1960, "the μ values are significantly lower in states of the North Eastern US, and higher in the Southern and Western states." They speculate that this may reflect greater isolation in the south and west. "Overall," as Alexander Bentley and Paul Ormerod conclude, "the diversification of naming practices across the US appears to be driven by an increase in innovation that drives further drift."[23]

This example nicely illustrates the usefulness of testing for neutrality before beginning any study that seeks selective causes. We find it particularly compelling because neutrality persisted despite the enormous demographic and cultural changes that occurred in the United States during the twentieth century. It also clarifies the difference between selection

[21] Hahn and Bentley. Drift as a mechanism for cultural change, S123.

[22] P. Barucca, J. Rocchi, E. Marinari, et al. 2015. Cross-correlations of American baby names. *Proceedings of the National Academy of Sciences USA* 112:7943–7.

[23] R. A. Bentley and P. Ormerod. 2012. Accelerated innovation and increased spatial diversity of US popular culture. *Advances in Complex Systems* 15:1150011.

as choice—the parents' careful but ultimately neutral deliberation—and selection as a driver of evolutionary change. Finally, it shows how neutral drift can create patterns of change that are observable at different scales of space and time.

Transience and time scales: Potsherds and archaeology

In 1995, Fraser Neiman introduced the neutral theory to archaeologists, giving full credit to its origin in population genetics: "[t]he models employed here were originally developed in population genetics to describe variation in neutral alleles. Only slight modifications are necessary for application to the horizontal, oblique, or vertical transmission of selectively neutral cultural variation."[24] Neiman applied Kimura's test, exactly as described above, to analyze stylistic change in ceramic cooking pots from southern Illinois dating from 200 BC to AD 800. The ceramics were obtained from nine sites occurring in five discrete site clusters or localities in the southern Illinois and Kaskaskia River valleys. In addition to testing for neutrality at each site, Neiman was also interested in the effects of trade or exchange between them. As with the baby names, answering this question required the analysis of change at two scales, global (the entire data set) and local (interactions between sites). After establishing that the stylistic variation was neutral at the global scale, Neiman turned his attention to the joint effects of two processes occurring at each site: drift and the exchange of pots between sites. In the tradition of neutral theory, he sought the simplest possible model to explore the dynamics.

Kimura's equation provided a way to test for drift at each site. In general, the smaller the site, the faster the rate of drift. As for exchange, Neiman chose geographic distance between sites as the simplest model. The sites varied in both size and distance: the average spacing between adjacent site clusters was about 130 km, with the closest pair separated by roughly 50 km. Neiman could have chosen to model mutation too, which in the archaeological context would have meant the appearance of new styles, but the data suggested that most new styles appearing at each site were introduced by exchange rather than local innovation. In any case, it would have been difficult to determine where and when innovations occurred. Provisionally excluding mutation (local innovation) simplified the model and increased its statistical power.

---

[24] F. D. Neiman. 1995. Stylistic variation in evolutionary perspective: Inferences from decorative diversity and inter-assemblage distance in Illinois woodland ceramic assemblages. *American Antiquity* 60:7–36.

To analyze the combined effects of both processes (drift and exchange), Neiman adapted a model from population genetics that was designed to track changes in gene frequencies as the combined result of drift and exchange among small, localized demes.[25] This method uses a matrix to track both migrations between demes and drift within each deme.[26] Neiman divided his samples into three temporal periods: Early, Middle and Late Woodland. For each period, he ran the model until the artifact distributions reached neutral equilibrium, and then observed the values for the exchange of pots (intergroup transmission). These values began low in the Early Woodland, rose to a peak during the Middle Woodland, and then fell to a new low in the Late Woodland. The model was validated by comparing its predictions against the frequency distribution of sherd variance at each site at different times.

These results, and Neiman's careful explanations, inspired a discourse on neutrality in archaeology that continues today. "Neiman pioneered virtually every technique used by archaeologists today to model and study cultural transmission," as Mark Madsen[27] noted in 2013. In archaeology as in genetics and ecology, the neutral theory was met with resistance: "most archaeologists shy away from explaining at least most cultural change as random."[28] Nonetheless, some archaeologists saw the advantages of neutrality as a null hypothesis for studying cultural evolution.[29] Moreover, as Neiman showed, if the underlying evolutionary process is neutral, one can draw inferences about interesting variation

---

[25] J. W. Wood. 1986. Convergence of genetic distances in a migration matrix model. *American Journal of Physical Anthropology* 71:209–19.

[26] "The magnitude of the covariance between two groups will scale inversely with their sizes and positively with the intergroup transmission rates. High levels of intergroup transmission, either directly or through an intermediate group, will mean that whatever departures from the initial frequency occur, they will be similar and in the same direction, hence the covariance between the groups will be high. On the other hand, if either or both group sizes are low, causing drift to play a stronger role, variant frequencies are less likely to depart from initial frequencies in a similar fashion, hence the covariance will be low." Neiman, Stylistic variation, 25.

[27] M. Madsen. 2012. Unbiased cultural transmission in time-averaged archaeological assemblage. *arXiv* 1204.2043.

[28] D. P. Braun. 1992. Why decorate a pot? Midwestern household pottery, 200 B.C.–A.D. 600. *Journal of Anthropological Archaeology* 10:360–97, p. 365.

[29] R. A. Bentley, M. W. Hahn, and S. J. Shennan. 2004. Random drift and culture change. *Proceedings of the Royal Society B* 271:1443–50; Neiman. Stylistic variation, 7–36; S. J. Shennan and J. R. Wilkinson. 2001. Ceramic style change and neutral evolution: A case study from Neolithic Europe. *American Antiquity* 66:577–94; A. Kandler and S. Shennan. 2013. A non-equilibrium neutral model for analysing cultural change. *Journal of Theoretical Biology* 330:18–25.

in population characteristics, such as population size and mutation rates over time.[30]

The limitations of the neutral theory for archaeological data were also noted. Some were not unique to archaeology. For example, balancing selection can lead to frequency distributions that appear to be neutral. And conversely, demographic processes can create distributions that appear to be the result of selection, but are actually neutral. James Steele and his colleagues[31] examined potsherds from Boazköy-Hattusa, the ancient capital of the Hittite empire and the largest Bronze Age settlement in Turkey. They found that while the frequency distribution of rim sherds did not in itself enable them to reject the null hypothesis of random copying, closer examination of the characteristics of these types revealed latent dimensions of functional variability (including ware type and bowl diameter) that had demonstrably been the subject of selective decision-making by the potters:

> Bowl shapes that were produced more often in plain coarse ware in the first phase tended to become more popular in the second phase. This is inconsistent with a neutral model and indicates that by the later phase, popularity of bowl types was associated with functionally-significant characteristics that had become subject to selective decision-making by the potters or their clients.[32]

Conversely, they observed that artifact distributions that fail the neutrality test may yet be neutral: "If, for instance, the innovation rate m has recently changed, or if N_e has recently changed, or if the population is incompletely mixing, then there may be an excess of variants at certain frequencies without any departure from neutrality in the underlying general process."

Conclusion

Versions of the neutral theory successively became among the most hotly debated topics in genetics, ecology, and the social sciences. The reasons for this are not hard to understand: aside from the prestige of Darwinian theory, any researcher who has spent time analyzing the functional role of a species, behavior, custom, or artifact will probably not be

[30] Ibid.

[31] J. Steele, C. Glatz, and A. Kandler. 2010. Ceramic diversity, random copying, and tests for selectivity in ceramic production. *Journal of Archaeological Science* 37:1348–58.

[32] Ibid., 1355.

delighted by the prospect that its prevalence may be the result of chance. But neutrality tests do more than serve as a null model. In population genetics, an array of statistical tests that distinguish neutral and selective effects, such as linkage disequilibrium and genetic hitchhiking,[33] provide increasingly fine-grained insights into change at the molecular level. In ecology, the neutral theory provoked much initial controversy, as it appears to abandon the role of ecology when modeling ecosystems. Subsequently, alternative neutral models were developed and tested on different taxa, along with hybrid approaches that aim to incorporate ecological dynamics while retaining Hubbell's concept of continuity of communities in space and time.[34] By incorporating short-term population fluctuations and environmental stochasticity, hybrid models "may serve as a minimalistic framework explaining fundamental static and dynamic characteristics of ecological communities."[35] Perhaps most relevant for the other social sciences are the models developed in archaeology, which have the advantage of building on decades of work in population genetics. But typical archaeological data consist of artifacts that have accumulated in the ground over varying time spans. These *time averaged* sequences (described as "accretional palimpsests" by Madsen[36]) are unlike data from population genetics or ecological communities, and may require variant approaches.

Selection and drift can occur in any evolutionary process, as populations are subject to the ravages of time. Neutral models offer a way to distinguish their effects by observing their outcome: countable haplotypes, names, species, or artifacts. But sometimes we are fortunate to see not only outcomes, but the processes that produce them. That can be especially important when several processes coevolve. In the next chapter, we explore this phenomenon in a study of the role of kinship systems in the transmission of language.

[33] Genetic hitchhiking occurs when an allele changes frequency not because it itself is under selection, but because it is near another variant—say, a gene—on the same chromosome that is undergoing a selective sweep.

[34] R. E. Ricklefs. 2006. The Unified Neutral Theory of Biodiversity: Do the numbers add up? *Ecology* 87: 1424–31.

[35] M. Kalyuzhny, R. Kadmon, and N. M. Shnerb. 2015. A neutral theory with environmental stochasticity explains static and dynamic properties of ecological communities. *Ecology Letters* 18: 572–80.

[36] Madsen, Unbiased cultural transmission, 1204.2043.

CHAPTER 4

Language and Kinship in Deep Time

> Language moves through time in a current of its own making.
> —*Edward Sapir*

In 1995, Richard Dawkins memorably described genes as a "river out of Eden,"[1] an unbroken connection between the first DNA molecules and every living organism. We are not accustomed to think of language in the same way. But we each speak a language that has been transmitted to us in an unbroken chain stretching back to the origin not of life, but of our species. Both the DNA molecule and language are made up of small units (codons for DNA, features for languages) that are subject to drift. Human DNA is transmitted by sexual reproduction, and as a by-product, so too are languages. Indeed these two processes, the transmission of DNA and of language, are not really separate. Every child inherits DNA from both parents and their first language from one or both. In the tribal world, the transmission of language and DNA is normally joined in a single channel, from parent to child. The continuation of that channel from one generation to the next depends on the cultural rules of kinship—marriage, residence, and inheritance—as well as occasional accidents of history. But language itself also protects those channels: shared languages help to define and connect groups of related individuals.

In previous chapters, we explored the insights into social behavior that can be gleaned from genetics. In this chapter, we will add language not as a separate topic, but as one that is intimately connected to kinship systems and thus to DNA. Along with language, we will also consider the transmission of culture. In most of tribal Indonesia, and especially in the islands east of Bali, there is usually a one-to-one correspondence between tribal languages and the decorative patterns woven into textiles used as clothing or for decoration. These functions of language as markers of identity and ancestral heritage are mostly absent from the modern world, although even today a visitor to out-of-the-way European villages may notice the correspondence of dialects with clothing styles and decorative motifs.

The idea that languages and lineages form channels that transmit a shared inheritance connecting the generations is elegantly expressed in the architecture of Sumbanese villages (Figure 4.1). Wherever possible they

---

[1] R. Dawkins. *River Out of Eden*. Basic Books, 1995.

Figure 4.1. Clan origin houses in a Sumbanese village. Credit: J. Stephen Lansing.

are built on hilltops. Each clan maintains a clan origin house, where they preserve the heirlooms that connect them to their male ancestors. Often only a few elders live in or near these houses. Most of the time, families live in temporary dwellings located near their gardens, and make brief visits to their origin village for important clan gatherings and feasts. The largest gatherings are death rituals for prominent men, marked by the erection of dolmen (single-chamber megalithic tombs) in front of the clan origin house (Figure 4.2).[2] At these mortuary rites, buffalo are slaughtered to feed the clan and their guests, and afterwards the horns of the buffalo are used to decorate the walls of the origin house. The size and magnificence of the horns and dolmen become a permanent public display of the clan's power to mobilize resources.

Thus, Sumbanese villages are not permanent residential communities, but sites where several clan origin houses are gathered together. Many clans have branches in several villages, and these clusters of villages will usually share a common language. Alongside or beneath their clan origin houses women often weave textiles, entwining symbols that represent neither her clan nor her village, but the cluster of villages that share a common language (Figure 4.3).[3] Worn as clothing, the textiles represent

---

[2] The accumulation of megaliths in such hilltop villages gives an impression of their age. But villages do not last forever, and a walk along the ridge lines will often lead to the remains of abandoned villages.

[3] Even today, weavers in East Sumba may be slaves.

Figure 4.2. Dolmen commemorating ancestry in the center of a ring of clan origin houses in a Sumbanese village. Credit: J. Stephen Lansing.

Figure 4.3. Weaving a sarong beside a Sumbanese clan origin house. Patterns are identified by the local language, and worn as clothing by those who share that language. Credit: J. Stephen Lansing.

a person's social identity. In this sense, shared language equates to shared culture, which is interpreted as an inheritance from the ancestors. The social world consists of unilineal descent groups or clans, which are gathered together in origin villages. Speech communities consist of the villages that contain the branches of the clans related by common descent. Because of this relationship, a shared language signals shared ancestry and the unbroken transmission of patrimony that defines social identity.

But the association of languages with descent groups does not last indefinitely, even on Sumba. Geneticists routinely test for correlations between genes, languages, and geography. The results are mixed. Gene trees can extend as far as one likes into the past. But language trees do not: unwritten languages change relatively quickly and persist in a reconstructable form for at most a few thousand years. If populations remain isolated for centuries, as in the rugged mountains of the Caucasus or New Guinea, all three (genes, languages, and geography) become strongly correlated. Otherwise, population movements and language drift cause these associations to weaken.

But perhaps correlations are not the best way to think about the relationship between language trees and gene trees. Intuitively, the two must somehow coevolve because a language exists only as long as it is retained in human communities.[4] The correlation is perfect while that association persists, but breaks down when people move or adopt a new language. In the past, such associations must have endured long enough for human languages to evolve their fundamental properties, because if communities switched languages too often (say, every generation), neither syntax nor semantics would have had time to mature.

Still, observing language use in human societies in the distant past would seem to be out of our reach without the use of a time machine. Linguists debate how long language families have persisted, but that is a different question from the social life of languages in the past. Can we learn anything about Sapir's "currents of language" in deep time?

Interestingly, the answer is yes. And the answer is in the trees. The key point is that at the community scale, the genetic and linguistic trees become joined. Human communities are fundamentally different from animal societies because they are organized by kinship systems, which are themselves adaptations. Every living person speaks at least one language, invariably inherited from one or both parents. If people sometimes marry

---

[4] In his 1980 letter *When is it coevolution?*, Daniel Janzen defines coevolution as "an evolutionary change in a trait of the individuals in one population in response to a trait of the individuals of a second population, followed by an evolutionary response by the second population to the change in the first." This focus on mechanism, rather than a simple correlation between traits, is key, as we shall see below. D. H. Janzen. 1980. When is it coevolution? *Evolution* 34:611–2.

into villages where a different language is spoken, their children will learn the language of the new community. We can see this in the paired trees, which record not only the language replacement, but the movement of the parent to a new community. By studying genetic and linguistic trees together, the recent movements of living people and their languages in the present day can be traced, and we can also reconstruct the travels of their ancestors. If we perform this analysis for whole communities (and even better, for groups of villages), a fine-grained picture emerges of the interaction of these two symbolic systems, language and kinship, as they play out from one generation to the next against the backdrop of genetic inheritance. We can discover the basic principles of their kinship system, the histories of their languages, and the extent and nature of contacts between communities, extending deep into the past. But to see all this, we need to understand how, where, and why the trees are joined.

The insight that led us to this way of thinking about language and social structure came from Sean Downey, an anthropologist who developed new methods to analyze our language data from the islands. His first contribution was to create a method to build language trees from our data.[5] From the very beginning of our research on the islands beyond Bali, we collected what are known as Swadesh word lists in each of the communities we studied. To do so, we recorded 200 specific words spoken in each local dialect. Linguists use the same set of words for all languages, allowing them to be directly compared. The recorded words are transcribed into the International Phonetic Alphabet (see examples on the book's website). The more similar are the words in two languages, the more closely those languages are related. When speakers of a language become separated, and seldom or no longer interact, drift takes over and the words gradually become different.

Online Resource: ALINE Model of Language Drift

The model of language change is available to explore in the book's online resources:

https://www.islandsoforder.com/aline.html

Downey's second insight was to suggest a way of joining the language trees with the gene trees by adapting a method used in ecology to analyze the evolution of host-parasite interactions. The classic example is the relationship between pocket gophers and lice that live in the gopher's fur.

---

[5] Peter Norquest and Brian Hallmark also made substantial contributions to this analysis.

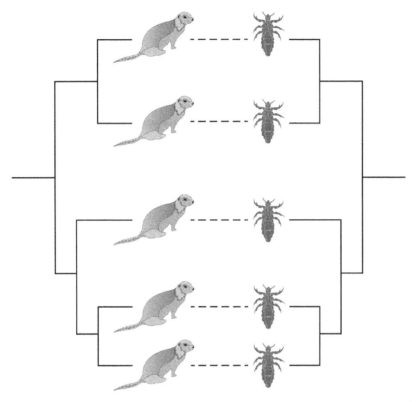

Figure 4.4. A cophylogeny of gophers and lice. In this illustration, each species of gopher hosts a unique species of louse. Credit: Yves Descatoire.

Most species of gophers have their very own species of lice (Figure 4.4). But from time to time, new species of both lice and gophers evolve—as the new species of gophers diverge, so do the lice they carry. So when, if ever, do the lice switch hosts? If we think of languages as parasites and human brains as their hosts, this is fundamentally the same question: when do languages switch to different brains?

Return to Wehali

In chapter 2, we encountered Wehali, the ancient matrilineal society of central Timor that inspired our model for the Austronesian expansion.[6] An intriguing feature of Wehali is its diversity of languages, cultures, and kinship systems. In the inner core of eight matrilineal communities,

---

[6] Learn more about the Wehali story in the video "Crossing the Wallace Line" on the book's website https://www.islandsoforder.com/exploring-austronesia.html.

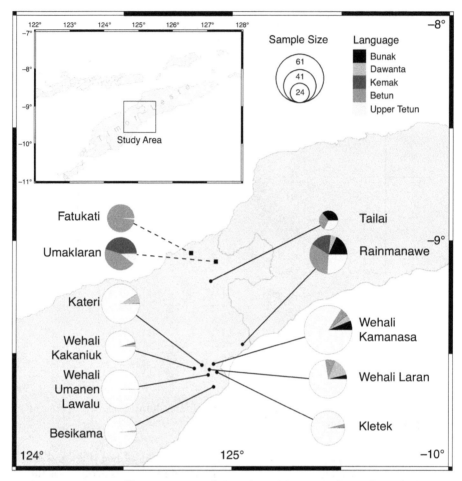

Figure 4.5. Map of languages in Wehali and neighbors. Pie charts show the languages spoken, scaled by sample size. Each of the 477 men sampled on Timor speaks one or more of five local languages belonging to two language families, Austronesian (Dawanta, Kemak, Betun, and Upper Tetun) and non-Austronesian Bunak. Credit: Authors.

four Austronesian languages are spoken (Figure 4.5). A non-Austronesian Papuan language is also spoken in several villages, and most (although not all) of the communities bordering the Wehali villages are patrilineal. Our first question is, can we adapt the cophylogeny approach developed by ecologists to make sense of this confusing picture?

The first step is to connect the languages to the people who speak them. A conventional statistical approach to this question might begin

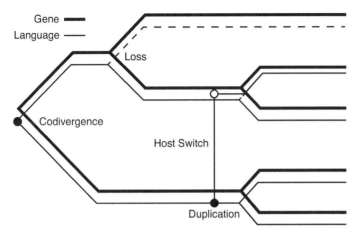

Figure 4.6. A cophylogeny of languages and genes, illustrating codivergence, loss, duplication, and host switching. Credit: Yves Descatoire.

by looking for patterns at the community scale. For example, is there an association between particular languages and either matrilineal or patrilineal kinship? As it turns out, the association is quite weak. But this question can be sharpened up by adopting a cophylogenetic perspective (Figure 4.6). To do so, we collected both genetic and language data from representative samples of men in each community. Our goal was to compare matrilineal and patrilineal pathways of language transmission for each individual. Although matrilineal descent can be inferred for both men and women from their maternally inherited mtDNA, patrilineal descent must be traced with the Y chromosome, carried only by men. Therefore, only men were sampled. In each community, we obtained DNA from a representative sample of men and also recorded the languages that each individual spoke.

Figure 4.7 compares two cophylogenies for all the men in our sample, a total of 477 men from 11 communities. Every man appears in both trees, together with exactly the same collection of male relatives. The difference between the two trees is that Figure 4.7A groups the men by their matrilineal kinship (a mitochondrial DNA tree), while Figure 4.7B groups them with their closest patrilineal relatives (a Y chromosome tree). The bands at the bottom of each phylogeny indicate the languages spoken by each individual. The top band shows all the languages that each man speaks: Upper Tetun, Bunak, etc. The lower band shows the single most common language shared by each group of very closely related men. Geneticists refer to such groups of close relatives as *clades*.

The association between clades and languages can come about in two ways. As mothers (or fathers) transmit a language to their children from

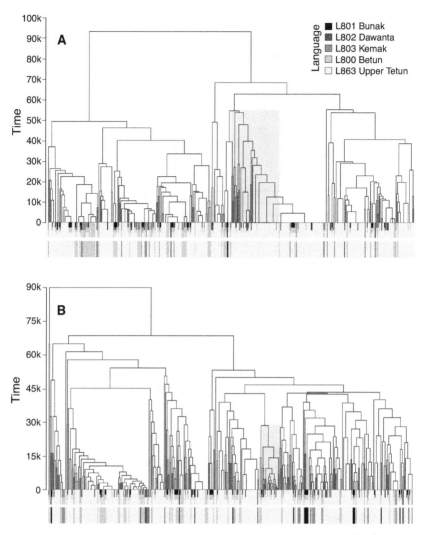

Figure 4.7. Language sharing in the cophylogenies of Timor. A: matrilineal descent traced by mtDNA haplotypes. B: patrilineal descent for exactly the same men, traced by Y chromosome haplotypes. Bands beneath the phylogenies show the languages spoken by each individual. The upper band shows all the languages spoken by each individual. The lower band shows the single language that is most widely shared by each group of close relatives. Overall, the size of these bands is larger for matrilineal descent groups. Credit: Authors.

one generation to the next, the number of descendants who inherit this language will increase. We can think of this as vertical transmission. Horizontal transmission is also possible: a group of closely related men may adopt a new language in as little as one generation. This will erase any trace of their former languages from the clade. In Wehali, our subjects confirmed that vertical transmission is dominant and is channeled by their marriage customs. This can be seen by comparing the two cophylogenies. Comparing the lower color bands in A (matrilines) and B (patrilines), there is a clear tendency for shared languages to extend over a broader swath of matrilineal clades than patrilineal clades. In other words, men who are more closely related on their mother's side are more likely to share a common language with their close matrilineal kin. Bear in mind that both trees—the mitochondrial (mtDNA) and the Y—contain exactly the same men and exactly the same languages. They differ only in the grouping of clades and the association of clades with languages. While most of the men probably inherited their language from both parents, in the long run, matrilineal transmission was stronger.

So should we infer that there is a general tendency to learn one's mother tongue? Or are these results caused by the matrilineal bias of kinship in Wehali? To find out, we needed a larger sample of patrilineal villages. Fortunately this was readily available on the neighboring island of Sumba. On Sumba, we surveyed 14 patrilocal villages where 12 Austronesian languages are spoken (Figure 4.8). In this way, our sample size grew to include 25 villages on two islands, with roughly equal numbers of matrilocal and patrilocal communities, speaking 17 languages belonging to two language families.

Cophylogenies of languages and genes

On Sumba, it is the custom for men to remain in their natal village for their entire lives and to marry women from neighboring villages, where a different language is often spoken. Young women begin to learn their husband's language when they marry and move to his village. Over time, women say, they tend to forget the language of their girlhood, as they learn their husband's language. All of the Sumbanese villages in our sample fit this pattern; they are patrilocal and monolingual. All but one language is spoken in only a single village, the exception being the Kambera language that is spoken in the two villages of Bilur Prangadu and Mbatakapidu. Consequently, on Sumba, there is a strong bias for patrilineal transmission of language, the mirror image of matrilineal Wehali. This

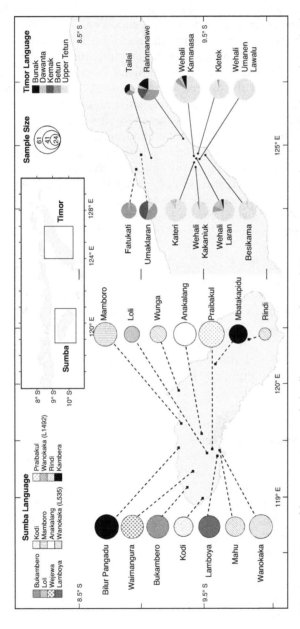

Figure 4.8. Comparison of languages in sampled villages on Sumba and Timor, eastern Indonesia. Pie charts show the languages spoken by sample size. Left: 14 patrilocal villages (■) on Sumba and the Austronesian languages spoken by the 505 sampled men. Right: 11 communities on Timor, including 9 matrilocal villages (●) and 2 patrilocal villages (■). Each of the 477 men sampled on Timor speaks one or more of five local languages belonging to two language families: Austronesian (Betun, Dawanta, Kemak, and Upper Tetun) or non-Austronesian Bunak. Multiple languages are typically spoken within communities on Timor, while only one language is usually spoken within a given community in Sumba. Credit: Authors.

contrast is clearly evident in the four cophylogenies[7] shown in Figure 4.9. Matrilineal relatedness is shown in the left-hand panels (A and C) and patrilineal relatedness in the right-hand panels (B and D). The bands at the bottom of each phylogeny indicate the languages spoken by each individual, including multiple shades if the individual is multilingual. The horizontal span of each segment in the band indicates the size of groups of individuals who both speak a common language and are closely related.

Consider Figure 4.9A, the top left panel showing the association of languages with matrilineal clades on Sumba. The tendency is for very small groups of men to share a common language with their closest matrilineal relatives. This could occur if brothers or cousins in the village often seek wives from the same neighboring village. A much stronger association is shown in Figure 4.9B, the association of languages with patrilineal clades. This pattern appears to be caused by the patrilocal kinship system, which prompts men to remain in their father's village, retain his language, and pass it on to his children. The overall statistical association between languages and both matrilineal and patrilineal clades is shown in Table 4.1. In patrilocal Sumba, a very strong association is seen between genes and languages on the Y chromosome ($Z = 67.7$, shown by gray shading in Table 4.1). In Timor, the two patrilocal villages show the same association, while in the matrilocal villages, languages follow matrilineal inheritance.

But these results merely describe the statistical association of clades and languages in the present generation. We have yet to take advantage of the information contained in the trees. The phylogenetic trees for mtDNA and Y chromosome (Figure 4.9) have roots in the very distant past, long before any conceivable relationship to the languages spoken today could have existed. Intriguingly, however, some information about the association between languages and genetic clades in the deep past is preserved by the branching points where languages either become attached to, or leave, genetic clades.

On both islands, a comparison of the topologies of the mtDNA and Y chromosome trees suggests that both patrilineal and matrilineal kinship systems can create durable channels for language transmission.

---

[7] The Y chromosome tree for patrilocal Sumba communities (Figure 4.9B) and the mtDNA tree for matrilocal Timor communities (Figure 4.9C) contain larger clades of related individuals who speak a common language (25 and 47 individuals in the largest clades on the Sumba Y chromosome tree and the Timor mtDNA tree) compared to the trees of the dispersing sex (12 and 26 individuals in the largest clades of Sumba mtDNA and Timor Y). Larger groups of individuals who have close genetic relationships and speak a common language are therefore found in the lineages of the nondispersing sex.

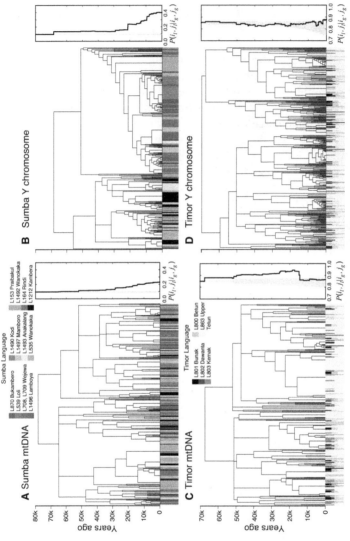

Figure 4.9. Comparison of gene-language cophylogenies of (A–B) patrilocal Sumba and (C–D) matrilocal Timor. Bands beneath the phylogenies show the languages spoken by each individual (monolingual in Sumba; sometimes multilingual in Timor). Plots to the right of each phylogeny show the probability of sharing a language l given that each pair of individuals are in the same genetic clade g at a given time in the past. Solid lines represent the observed metric, with shaded bands indicating the results of random permutations of the linguistic data. Higher probabilities that close genetic relatives share a language, compared with random expectations, are observed for (B) Sumba Y and (C) Timor mtDNA at all time periods. For (A) Sumba mtDNA and (D) Timor Y, probabilities are only higher than random cases until ~65,000 and ~1,500 years ago, respectively. Credit: Authors.

Table 4.1

Z scores of gene-language associations. Comparing within each group of villages (columns), stronger gene-language associations (shading) are found for the Y chromosome in all Sumba villages, mtDNA in all Timor villages, mtDNA in matrilocal Timor villages, and Y chromosome in patrilocal Timor villages.

	Sumba			Timor		
	All	*Matrilocal*	*Patrilocal*	*All*	*Matrilocal*	*Patrilocal*
mtDNA	8.32	—	8.32	5.43	6.62	1.01
Y	67.7	—	67.7	3.85	4.04	2.67

The vertical transmission of languages along genetic clades would tend to produce persistent speech communities. To test this hypothesis, we define a probability[8] that a pair of individuals (i, j) share a common language l, given that they belong to the same genetic clade g at some time in the past t. The results of this calculation are shown in the plots located to the right of each tree in Figure 4.9, with solid lines representing the observed data and shaded regions indicating the range of probabilities seen when languages are shuffled randomly among samples. As we track back through time (i.e., upwards along the branches and deeper into the past), men who are related on their fathers' side in the patrilocal villages of Sumba are consistently more likely to speak a common language compared to random cases. This tendency is stronger along patrilines (Figure 4.9B) than matrilines (Figure 4.9A). Conversely, in mostly matrilocal Timor, men are more likely to share a common language with their close matrilineal kin (Figure 4.9C–D).[9]

But as the trees also show, the association between languages and clades does not last forever. In each generation, an opportunity exists to weaken or scramble the correlation between genetic and language lineages through host switching, which occurs when children learn a different language than one of their parents. This can lead to three outcomes. In

---

[8] Formally, $P(i_l, j_l | i_{g(t)}, j_{g(t)})$. For details, see J. S. Lansing, C. Abundo, G. S. Jacobs, et al. 2017. Kinship structures create persistent channels for language transmission. *Proceedings of the National Academy of Sciences USA* 114:12910–5.

[9] The initial similarity to the random distribution ($\sim 15{,}000$ years ago to present) seen in the language sharing probabilities of Timor men related through their mothers (Figure 4.9C) is caused by the presence of two patrilocal villages in otherwise matrilocal Timor, together with the simultaneous branching of many individuals (polytomies), reflecting both demographic history (rapid periods of growth) and a lack of resolution in our genetic markers, especially toward the present. However, in comparison, Timorese related through their fathers (Figure 4.9D) have language sharing probabilities that are within the random range for almost all generations (oldest generation to $\sim 1{,}500$ years ago).

the absence of host switching, the emergence of new languages (branching in the language tree) can introduce persistent clades of related individuals who speak a common language, thus creating extensive correlation between the gene and language trees. When host switching occurs rarely, some correlation remains, but frequent host switching and language losses quickly break down the correlation. So is host switching a random process, like drift? Does it depend on large-scale population movements like migrations? Does it vary between the islands or between matrilineal and patrilineal communities?

To find out, we can estimate host switching probabilities for the observed gene-language tree associations in each of the cophylogenies. Assuming that both gene and language trees are accurately reconstructed, we can generate a one-to-one mapping of the branching points in the gene and language trees and therefore describe how the languages were transmitted.[10] A plausible model, shown in Figure 4.10, predicts the languages spoken at every branch in the gene tree in each generation t. This stochastic simulation is run forward in time over the trees using two rules. First, where the language tree branches into two daughter languages, all genetic clades that speak the ancestral language are randomly assigned to one of the daughter languages. Second, in each generation, there is a probability α that a given clade switches to a new language chosen from among the languages that exist in the population at that time. That is, a proportion of lineages α switches to a new language at each generation. These two rules provide a simple model of language diversification and host switching that is sufficient to reconstruct patterns of language sharing observed in the present.[11]

From this model, a mapping of the languages spoken at present by every individual in the gene tree (Figure 4.10C) can be generated. To examine

[10] P. Legendre, Y. Desdevises, and E. Bazin. 2002. A statistical test for host-parasite coevolution. *Systematic Biology* 51:217–34; K. Hunley. 2015. Reassessment of global gene-language coevolution. *Proceedings of the National Academy of Sciences USA* 112:1919–20.

[11] This stochastic model predicts language transmission along the branches of the gene tree run forward in time starting from the gene tree root (t_1 in Figure 4.10C). The following process is performed at each generation (taken to be 30 years): i) determine the genetic clades and languages at generation t (e.g., two genetic clades and languages L_{12}^* and L_3 at t_2 in Figure 4.10B–C); ii) if a language branches at generation t, each genetic branch that carried the ancestral language at $t-1$ is randomly assigned one of the two new languages (e.g., at t_2 in Figure 4.10C, L_{12}^* is assigned to the first branch and L_3 to the second branch); iii) each genetic branch then switches to a different language within the current pool of languages with a probability α (e.g., genetic branch G_2 switches to language L_3 at t_4). Variants of this gene-language coevolution model yield qualitatively similar results.

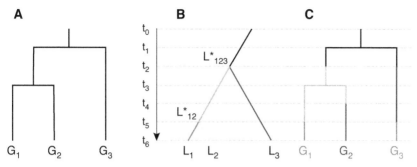

Figure 4.10. Language switching in a gene tree conditioned on a language tree. Shown in (A) is an unannotated gene tree, (B) a language tree, and (C) an example of a simulated language-annotated gene tree. Language diversifies i) during language branching (L^*_{123} splits into L^*_{12} and L_3 at t_2), and ii) during host switching, as in the branches leading to G_2 (language switch at t_4) and G_3 (language switch at t_5). Credit: Cheryl Abundo.

the extent of congruence between the gene and language trees given the mappings simulated from the model, we used a standard cophylogeny statistical test called *ParaFit*.[12] This test was repeated for many gene-language simulations across the range $\alpha = [0, 100]$, where an α of zero indicates no language switching and $\alpha = 100$ means all lineages switch languages every generation. The average Z score of each α value is then compared with the Z score of the observed data to discover the frequency of language switching that best fits the data.

In this way, we analyzed eight different cophylogenies as shown in Figure 4.11. In all eight cases, host switching rates converge to 0.3%–1% per generation for all eight trees. This translates to a ~50% probability that a single host switch event will occur within a clade (i.e., the "half life" of a language on a lineage) every $1,700$–$5,750$ years, with near certainty of a host switch event occurring over much longer time frames. At rates only slightly faster than 0.5%, the association between languages and clades quickly becomes random. Only models with very low rates of host switching ($< 0.5\%$) can generate cophylogenies with strengths of association like those seen in the observed data.

The remarkable similarity of host switching rates in all eight cophylogenies suggests that they were produced by the same process: kinship rules, not migrations or conquests. While host switching rates are similar

[12] S. Dray and P. Legendre. 2008. Testing the species traits-environment relationships: The fourth-corner problem revisited. *Ecology* 89:3400–12; Legendre, Desdevises, and Bazin, A statistical test.

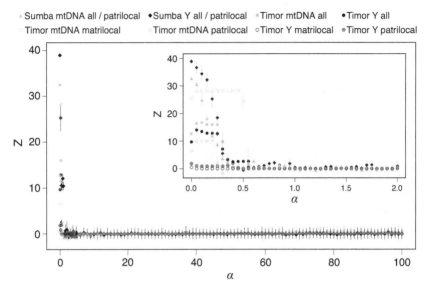

Figure 4.11. Language switching rates. The Z score measure of association between gene and language phylogenies on Sumba and Timor for different language switching rates α. All cases independently converge and abruptly lose gene-language associations, behaving similarly to randomized cases, when the language switch rate exceeds ~0.5% per generation. Credit: Cheryl Abundo.

for all cases, there is also a clear tendency for rates to be higher for men in matrilineal Wehali and women in patrilineal Sumba. In both cases, the sex that leaves home to marry is more likely to pass on a different language to their children. Thus we see two patterns in the 25 communities included in our study. Core groups of close relatives tend to stay together for many generations. However, they also stay in contact with neighboring groups with whom they intermarry. Perhaps the most surprising result is the inference that clades can retain a shared language for longer periods of time than any single language exists, by groups of people replacing one language with another in concert.

The implications of host switching

We pause here to consider the implications of these results. The mtDNA and Y chromosome phylogenies tell us how closely the men in these communities are related, but provide no information about how those relationships came to be. Joining the gene and language trees neatly solves this problem. For each individual in our sample, we construct two cophylogenies of language and genetic inheritance for matrilineal and patrilineal markers, respectively. If there were no migration between villages, the signal of association in these cophylogenies would be identical, because men would learn the same language spoken by both parents. However, if people sometimes marry into villages where a different language is spoken, the gene and language phylogenies will diverge. If many languages are spoken within a geographical region, and rules of post-marital residence encourage sustained directional population movement between speech communities, then languages should be channeled along uniparental lines. We find strong evidence for this pattern on both Sumba and Timor.

While language and kinship are usually treated as unrelated topics, the cophylogenies suggest that in tribal societies their interaction is better understood as a dynamical process, which largely determines the relationship between languages and social groups. Kinship rules channel languages along clades into communities where some languages are learned and spoken, while others are quickly forgotten. The low rates of host switching suggest that this process of channeling may continue for as long as the kinship system remains unchanged, perhaps even longer than most villages or languages exist. To gain deeper insight into this process, we can shift our gaze from the trees to the villages. What is the relationship between the channels created by kinship practices, the transmission of languages, and the genetic composition of the villages?

Kinship and population genetic structure

The first question is how kinship systems affect the movements of men and women between villages. By taking advantage of our genetic data, this question can be easily answered. For this purpose, we used an Isolation with Migration (IM) coalescent model[13] to capture the genetic consequences of differences in female and male migration rates. How are these rates affected by cultural rules for postmarital residence: i) village endogamy, ii) ambi– or neolocality, iii) patrilocality, and iv) matrilocality?

The IM model describes a single panmictic population of size aN that splits into n subpopulations of size N at time $2N\tau$ generations in the past. Migration occurs at a rate m between subpopulations. To assess the effects of postmarital residence practices on mtDNA and the Y chromosome, we distinguish between the migration rates of women (mtDNA) and men (Y chromosome), and run the model separately for each group.

---

[13] H. M. Wilkinson-Herbots. 2008. The distribution of the coalescence time and the number of pairwise nucleotide differences in the "Isolation with Migration" model. *Theoretical Population Biology* 73:277–88.

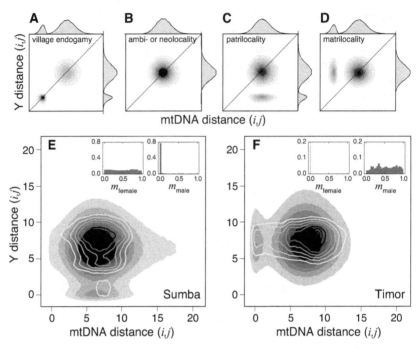

Figure 4.12. Genetic structure from an IM model with migration influenced by kinship practices. To identify the theoretical role of kinship practices on genetic diversity, the IM model was run for populations that are (A) endogamous, (B) ambi– or neolocal, (C) patrilocal, and (D) matrilocal. (E) Close correspondence of the IM model (shaded contours) with observed data (contour lines) for patrilocal Sumba using $N = 298$, $n = 41$, $m_{female} = 0.73$, $m_{male} = 9 \times 10^{-5}$, $\tau = 5.49$, and $a = 0.024$. (F) Close correspondence of the IM model with matrilocal Timor using $N = 264$, $n = 97$, $m_{female} = 1 \times 10^{-5}$, $m_{male} = 0.25$, $\tau = 6.71$, and $a = 0.06$. The insets (E) and (F) show the posterior distributions of migration rates for the majority kinship system in Sumba and Timor based on three billion samples drawn from prior distributions of all IM model parameters. Credit: Cheryl Abundo.

Both runs have the same values of N, n, a, τ, and mutation rate μ, but can have different migration rates m_{female} and m_{male}. For example, matrilocal kinship systems lead to greater migration of men, such that $m_{male} > m_{female}$. The converse is true for patrilocal systems. A cultural preference for endogamy is reflected by low migration rates for both m_{female} and m_{male}.

Figure 4.12A–D shows typical outputs from this model in the form of paired genetic distances. Here each individual is paired with every other individual, and the pairs are represented by single points corresponding to how closely they are related to one another on both mtDNA (matriline) and the Y (patriline). In the simplest cases (endogamy, and ambi– or neolocality), there is no bias toward matri– or patrilineal relatedness (Figure 4.12A–B). Consequently, given equal mutation rates, the distribution of pairwise mtDNA and Y distances lies on the one-to-one correlation line, at a distance from the origin that depends on demographic parameters. If both females and males marry and reside within their natal villages (Figure 4.12A), we see two clusters of pairwise genetic distances: i) near the origin, a cluster of closely related kin residing in the same village, and ii) a larger cluster consisting of individuals living in different villages, and thus less closely related. When there is no gender bias in dispersal and both sexes move frequently, the distributions of genetic distances for mtDNA and the Y are similar (Figure 4.12B). Introducing a matri– or patrilocal bias in marriage customs shifts the pairwise distance plot toward one or the other axis (Figure 4.12C–D). In patrilocal villages, where men remain in their natal villages while women move to marry, pairs of men remain closely related on their Y, but not their mtDNA (Figure 4.12C). The opposite holds for matrilocal villages where women remain in their natal villages and men move after marrying (Figure 4.12D).

These predictions are borne out in the genetic data from Sumba (Figure 4.12E) and Timor (Figure 4.12F). By plotting the pairwise genetic distance, we can see how closely individuals are related on their mtDNA (matriline) and Y chromosome (patriline). On Sumba, the cluster of small Y distances (close patrilineal kin) is more distinct, comprising 11% of all pairs, as compared to the faint cluster of small mtDNA distances (close matrilineal kin) with only 5.1% of pairs. On the other hand, on Timor, the more pronounced cluster is of small mtDNA distances comprising 8.3% of all pairs, in contrast to the cluster of small Y distances with 7.6% of pairs. Comparing the two islands (with more detail shown in Figure 4.13A–B), there is a clear tendency for closer patrilineal relatedness on Sumba, while on Timor we see the opposite pattern of closer matrilineal relatedness.

How much time is needed for these patterns to emerge, assuming a constant sex bias in migration rates? To find out, we used approximate Bayesian computation (ABC) rejection sampling[14] to assess which IM model parameters closely match the data. This method returns the posterior distribution of female and male migration rates that best fit the observed data (Figure 4.12E–F insets). The migration rate of the rarely

---

[14] M. A. Beaumont, W. Zhang, and D. J. Balding. 2002. Approximate Bayesian computation in population genetics. *Genetics* 162:2025–35.

dispersing sex is tightly constrained, while the migration rate of the other sex is not.

Online Resource: Sex-Biased Migration

The model of sex-biased migration is available to explore in the book's online resources:

https://www.islandsoforder.com/sex-biased-migration.html

Kinship and language transmission

Thus, the genetic evidence suggests that sex-biased migration rates on both islands are consistent with the observed kinship rules and have persisted for many generations. Do these kinship practices sustain the association between languages and genes, and in so doing create persistent speech communities? To find out, we analyzed pairwise distance plots weighted by $d_l(i,j) = \cos(l_i, l_j)$, describing the degree of language sharing between individuals. A pair of individuals (i, j) speaking exactly the same set of languages l_i and l_j will have $d_l(i,j) = 1$. If they speak some, but not all, languages in common, then $0 < d_l(i,j) < 1$. If they speak no languages in common, $d_l(i,j) = 0$. If languages are transmitted along uniparental clades, the genetic distances between pairs of individuals who speak a common language should enhance, and further reveal, clusters that reflect the expected sex-biased migration patterns. To test this inference, we compare the pairwise distance plots weighted by degree of language sharing (Figure 4.13C–D) with the unweighted plots for all pairs of individuals (Figure 4.13A–B).

On Sumba, the signal of patrilocality is strongly enhanced (reaching 25% of sampled pairs) when only individuals who share a language are considered (compare Figure 4.13C with 4.13A). All of these villages are monolingual, and all but one language is found in just one village, the exception being Kambera, which is spoken in the two villages of Bilur Prangadu and Mbatakapidu. Consequently, most paired individuals who speak the same language are from the same village. While this may seem to be a limitation in the data, it is actually ideal for the purpose of distinguishing whether each man inherits his language from his patriline, matriline, or both. Comparing Figure 4.13A (all pairs of Sumba men) with Figure 4.13C (only men who share the same language), there is a very strong trend for the male children of women marrying into a community to learn the language of that community.

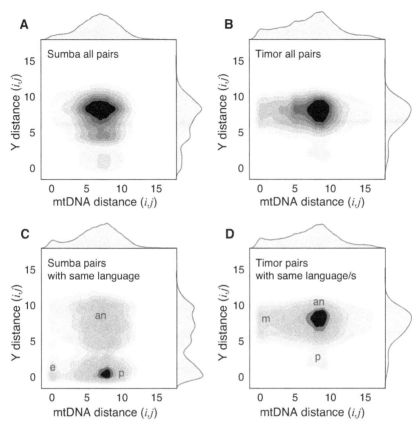

Figure 4.13. Genetic distances on Sumba (left, A and C) and Timor (right, B and D) between all pairs of individuals (A–B) and only between individuals who speak a common language (C–D). Conditioning on language sharing reveals three distinct clusters in the Sumba data (C), showing evidence of village endogamy (e), ambi- or neolocality (an), and patrilocality (p). For Timor, the high degree of multilinguality means that most pairs of individuals in B are also included in D. B and D are dominated by matrilocality (m) and ambi- or neolocality (an), with a small patrilocal cluster (p). Credit: Cheryl Abundo and Guy Jacobs.

Timor differs from Sumba in two key ways: our sample includes a mix of nine matrilocal and two patrilocal villages, and multilinguality is common. Consequently, resulting pairwise distances show a more complex pattern of three clusters (Figure 4.13D):

1. A matrilocal cluster (**m**), comprising pairs of men who are closely related on the matriline;

2. A weak patrilocal cluster (**p**), comprising pairs of men closely related on the patriline due to the two patrilocal villages in the sample; and

3. A large ambi– or neolocal cluster (**an**), comprising pairs of men who are less closely related on both matriline and patriline. Because the Timor sample includes both matrilocal and patrilocal communities, the unbiased ambi– or neolocal cluster is the most pronounced across the entire sample of villages on Timor.

Importantly, conditioning on the number of languages shared enlarges the matrilocal cluster by an additional 0.8% (compare Figure 4.13D with Figure 4.13B), the converse pattern of Sumba.

Thus, the extent of language sharing further highlights the expected sex-biased migration patterns observed from genetic distances. These results are consistent with the overall statistical cophylogenetic association between languages and either matrilineal or patrilineal genetic clades (see Table 4.1). In patrilocal Sumba, a far stronger association is seen between genes and languages for the Y chromosome ($Z = 67.7$). A similar pattern ($Z = 2.67$ for the Y chromosome) exists for the two patrilocal villages in the Timor sample, while the opposite pattern ($Z = 6.62$ for mtDNA) is found for the matrilocal villages of Timor.

Language and kinship in deep time

It is customary to attach an estimate of time, based on the molecular clock, to the branches of gene trees. Linguists use a similar method, called *glottochronology*, to estimate the age of language trees. Our first attempts at joining the trees to construct cophylogenies startled us: many branches appeared to show associations stretching back not for a few centuries, as we expected, but impossibly far into the Pleistocene. We struggled to make sense of this initial impression. Mathematical models helped, but the calculations only started to make sense when we stopped puzzling over dates and focused instead on speech communities. Historical linguistics focuses on language change, like the emergence of the Romance languages from Latin. Languages, not their speakers, are the target of analysis. For us, the priority is reversed. We discovered that groups of related individuals often share a common language over very long time spans. But while we know which languages they speak today, we do not know—at least for sure—which languages they spoke in the past. What we observe are the speech communities, not the languages.

A thought exercise may help to clarify this distinction. Closely related individuals are more likely to speak the same language—for instance, Bunak or Upper Tetun. But what would happen if we changed these languages overnight? It turns out that it depends on how you change them. Imagine that all the Upper Tetun speakers across our communities begin speaking Portuguese, while the Betun speakers adopt Dutch. Kemak is replaced with Italian, Dawanta with Romanian, and the non-Austronesian language Bunak with the non–Indo-European language, Turkish. The point is that *none of our results would change.*[15] The closely related individuals who once spoke Upper Tetun are now simply closely related individuals who speak French. The same outcome holds for everyone else. The crucial insight is that as long as the speech communities adopt their new language as a group, the gene-language associations persist—even if the actual languages they speak have changed.

But what if instead people switch languages randomly—this person speaks French, that one Spanish, based—say—on the color of the t-shirt they happen to be wearing today? The association between closely related individuals and the shared language(s) they once spoke would disappear. This is the key take-home message: it is communities of language speakers that persist through time, while the languages themselves are always changing. Consequently, correlations between languages and genes can persist for what seems like impossibly deep times into the past, sometimes long before the languages spoken today even arose.

Zooming in to the community scale

So far our analysis has focused on discovering general principles that are apparent at the broadest scale, the comparison of 25 villages on two islands. But now that these principles are in hand, can we zoom in to the village scale, and observe their functional effects?

Consider the Sumbanese region of Kodi, defined by the Kodi language and located on the west coast. We sampled two communities, located about a kilometer apart, which are considered to be the same origin village. As Figure 4.14 shows, Kodi is strongly patrilineal. The samples were drawn from 45 men who belong to half a dozen patrilineal clans, nearly all of which show the same genetic pattern: very close recent ancestry on the father's side, together with a broad range of ancestry on the mother's side. A small group of men are not closely related on either the father's or mother's side, and show up as a faint cloud of points in the

[15] Strictly, the analysis also depends on the phylogeny of the languages. Because the Austronesian languages of Timor have different tree relationships to the Indo-European Romance languages of southern Europe, the results would actually differ sightly. However, the key message of this thought exercise still holds.

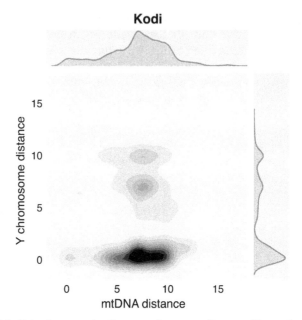

Figure 4.14. Pairwise genetic distances in two adjacent villages in the Kodi-speaking region of West Sumba. These men belong to very closely related Y chromosome haplotypes, but their matrilineal relatedness is much more diverse. The data include 21 samples from Tossi (Wuruhumbu) (location 9° 34.951 S, 118° 57.579 E) and 22 samples from Mbukubani (Atedalo) (about one km distant, but considered to be the same village, location 9° 34.125 S, 118° 56.609 E). The clans are Ndelo, Watu Pakade, Lamete, Mbahewa, Mbukabade, and Mbarada. Credit: Guy Jacobs.

center of the figure. Within that group, one would need to go back in time to an era long before the Austronesians arrived in Sumba to discover their most recent common ancestors on the patriline or matriline.

But not all Sumbanese villages are made up of very closely related patrilineal clades. Near the center of the island, where the Anakalang language is spoken, we sampled another half dozen clans in the origin village of Pasunga (Figure 4.15). The pairwise distance plot here shows two patterns. Roughly a third of the men are close patrilineal relatives, like nearly all of the Kodi clans. The largest group is not closely related on either side. However, this group shows a wider range of variation among patrilineal clades than the analogous cluster in Kodi. Importantly, there is a small cluster near the origin made up of men who are very closely related on both sides. Just above them, a larger group is more closely related on the mother's side. The pattern shown in Figure 4.15 indicates that the men of Pasunga must have been born into a different clan and a

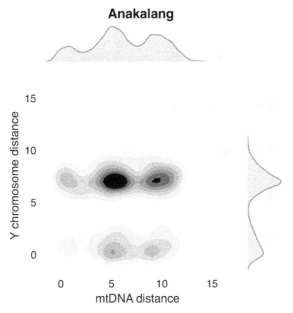

Figure 4.15. Pairwise genetic distances in the origin village of Pasunga, in the Anakalang region of central Sumba. This village is monolingual. The clans are Galubua, Preirita, and Laikaruda. Credit: Guy Jacobs.

different village, and married into Pasunga. If they held lower status than their wives, these marriages would have strengthened the alliance ties of the patrilineal clans of Pasunga. The data also suggest that the village is largely composed of two distinct groups of men; this is visible in the four most prominent clusters.

On Timor, the hamlet of Umanen Lawalu is one of the core communities in the ancient matrilineal society of Wehali (Figure 4.16). Roughly half the men are closely related to one another by shared matrilineal descent. The community is monolingual in Tetun, the dominant language of Wehali. To interpret the relationship between language and kinship in these villages, we need to take into account some important differences in the social structure of Wehali compared to Sumba. On Sumba, the largest political units are patrilineal clans. On Timor, more complex forms of social organization exist. In Wehali, the largest political structure is an alliance of eight matrilocal villages. As on Sumba, each village comprises several clans, which differ from Sumba in that they are matrilineal. On both islands, most villages actually comprise smaller residential units that we can translate as hamlets. The Timorese ethnographer Tom Therik notes that Tetun speakers have no specific term for "domain" or

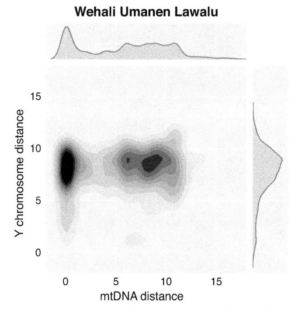

Figure 4.16. Pairwise distances for 48 men from the matrilocal Timorese village of Umanen Lawalu, one of the core villages of Wehali, comprising 535 families. The elected head of the village is a woman. Location 9° 35.682 S, 124° 53.658 E. Credit: Guy Jacobs.

"kingdom."[16] They believe that Wehali is the center of the earth, or at least of Timor. In primordial times the earth was covered by water. The first dry land emerged as a hill in Wehali, formed from the umbilical cord of an ancestress.[17] In inner Wehali, marriages between exogamous matrilineal clans often replicate alliances based on origin myths. This may help to explain the absence of a strong matrilineal signal in Laran, the innermost community in Wehali (Figure 4.17). In Laran, men who are closely related matrilineally are in the minority. A possible explanation is that Laran is regarded as the oldest and most central village in Wehali. Consequently the other communities seek to marry into Laran. A marriage system that imports men from many matrilocal communities could produce the observed distribution of pairwise distances, a genetic pattern that differs from the signal created by bilateral alliances between clans, as seen in Anakalang.

[16] Tom Therik was born and raised on the small island of Roti, just to the west of Timor.

[17] T. Therik. *Wehali, The Female Land: Traditions of a Timorese Ritual Centre.* Pandanus Books, Australian National University, 2004.

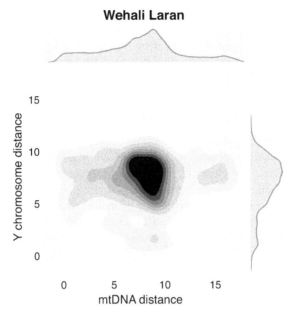

Figure 4.17. Pairwise genetic distances in the Timorese village of Wehali Laran, a matrilocal and multilingual community located in the center of the Wehali region. With 186 families, the total population was 686 inhabitants in 2008. Credit: Guy Jacobs.

Our final example is the village of Raimanawe, where four languages are spoken. This village is located 20 km east of Wehali, very close to the border with the independent nation of East Timor. Nearly all of the inhabitants are refugees from East Timor, most of whom arrived around 1999. They say that they came from two regencies (*kabupaten*) in East Timor: Suai Kofalima and Ainaru. We sampled 47 men from a total population of 915 households, most of whom belong to one of eight clans. In addition to the matrilineal signal, new detail appears: it is largely made up of many small clusters of men who share a common language. Within each cluster, pairs of men are related to one another at approximately the same genetic distance. Here we find a dramatic example of multiple recent migrations, but so far, little evidence of host switching.

Conclusion

Previous correlational studies of gene and language trees have focused on the effects of drift, geography (isolation by distance), and large-scale population movements to explain patterns of correlation at different scales of space and time.[18] In contrast, our cophylogenetic analysis of communities clarifies how language transmission is actively channeled and continually renewed by kinship practices, creating the observed patterns of language diversity and leaving a strong signal in population genetic structure. Analysis of gene-language cophylogenies and pairwise distances suggests that this channeling process usually persists long enough to be regarded as the norm. The Timor data make this point particularly clearly—beneath the surface of community scale variation in language diversity created by recent population movements, enduring associations of language with uniparental clades still persist. In the 25 communities included in our study, core groups of close relatives must have stayed together for many generations while also remaining in contact with neighboring groups with whom they intermarried. In this way, kinship systems directly shaped the language phylogeny over time: consistently following a postmarital residence rule turned social communities into speech communities. The resulting patterns are not random, but islands of order.

[18] J. S. Lansing, M. P. Cox, S. S. Downey, et al. 2007. Coevolution of languages and genes on the island of Sumba, eastern Indonesia. *Proceedings of the National Academy of Sciences USA* 104:16022–6; N. Creanza, M. Ruhlen, T. J. Pemberton, et al. 2015. A comparison of worldwide phonemic and genetic variation in human populations. *Proceedings of the National Academy of Sciences USA* 112:1265–72; F. M. Jordan, R. D. Gray, S. J. Greenhill, and R. Mace. 2009. Matrilocal residence is ancestral in Austronesian societies. *Proceedings of the Royal Society B* 276:1957–64.

~~~~~~

CHAPTER 5

Islands of Cooperation

> Felix qui potuit cognoscere causas, fortunatus et ille deos qui nouit agrestis.[1]
>
> —*Virgil*, Georgics

Prelude: How Bali became Bali

In the traditional tribal societies of Indonesia, the social world is entirely and exclusively organized by kinship systems, of the kind that we have explored in previous chapters. The largest social groups were villages. Today on many islands these villages are still known by the ancient Austronesian term *wanua*.[2] But on the islands of Bali and Java, kingdoms arose in the late first millennium AD. As their power grew, they gradually broke down the autonomy of the wanua. This process can be dated with some precision because the early kings frequently issued charters to the villages within their domains, which were etched onto metal plates in a successful effort to ensure their preservation. The Javanese royal inscriptions show that before AD 922 the wanua were governed by councils of village elders. By the late twelfth century, these councils disappeared from the village charters, and by the fourteenth century the term wanua itself vanished, replaced by named collections of hamlets that were directly under royal rule.

The earliest royal inscriptions by Balinese kings, dating from the ninth century AD, also refer to villages as wanua governed by councils of elders. But by the twelfth century, events on Bali began to take a different course. In Bali, wanua never disappeared from the slopes high on the volcanoes, where irrigated rice is not grown.[3] But by the fourteenth century, as the kingdoms of Java were entering into an imperial phase, in Bali the era of royal inscriptions came to an end, and Balinese kingdoms disintegrated into feudalities.[4]

---

[1] "It is well for one to understand causes, fortunate also to comprehend the gods of the countryside." Publius Vergilius Maro, *Georgics*, 2.490.

[2] The asterisk denotes a reconstructed ancestral word form.

[3] Instead, these highland Balinese wanua became known as ancestral villages, from which daughter communities arose.

[4] For a summary of this history, see J. S. Lansing, *Perfect Order: Recognizing Complexity in Bali*. Princeton University Press, 2006, pp. 42–54.

What caused the fragmentation and loss of power of the Balinese kings? And why did the ancient customs of Austronesian wanua survive only in the highlands? In the chapters to follow, we offer an explanation. In the regions of Bali where irrigated rice could be successfully grown, new forms of social institutions emerged. They were based neither on kinship ties, nor on the power of feudal lords. Instead, they created durable and effective channels of cooperation that extended across whole landscapes of villages and princedoms, and continue to function today. To understand how this was possible, we turn our attention from the workings of kinship to the emergence of cooperation.

Introduction

Ever since Émile Durkheim's 1893 analysis of social solidarity, *De la division du travail social*,[5] the question of how and why cooperation persists in societies has been central to the social and evolutionary sciences. In 1944, the study of cooperation acquired a mathematical foundation with the publication of the *Theory of Games and Economic Behavior* by John von Neumann and Oskar Morgenstern.[6] But in classical game theory, the key question was not the emergence of cooperation, but the analysis of competitive strategies. In 1972, cooperation itself became a theoretical question when evolutionary theorists John Maynard Smith and George Price showed how game theory could be adapted to analyze the evolution of various forms of cooperation, including altruism.[7] When are players satisfied with a tie instead of a competitive victory?

Later, when researchers began to focus on the relationship between human institutions and the environment, once again cooperation became the focal question because it effectively determines the sustainability of what came to be called *common pool resources*, or *the commons*. Most recently, the idea of planetary boundaries—that society as a whole has entered an era in which human-environmental interactions are increasingly perilous—has intensified interest in these questions.[8]

[5] É. Durkheim. *De la division du travail social: Étude sur l'organisation des sociétés supérieures*. Alcan, 1893.

[6] J. von Neumann and O. Morgenstern. *Theory of Games and Economic Behavior*. Princeton University Press, 1944.

[7] J. Maynard-Smith and G. R. Price. 1973. The logic of animal conflict. *Nature* 246:15–8.

[8] In 2007, a formal standing program in the *Dynamics of Coupled Natural and Human Systems* was created by the US National Science Foundation. A. Marina, H. Asbjornsen, L. A. Baker, et al. 2011. Research on Coupled Human and Natural Systems (CHANS): Approach, challenges, and strategies. *Bulletin of the Ecological Society of America* 92:218–28.

This is the first of three chapters that focus on the emergence of cooperation in the context of human-environmental interactions. All three chapters draw on the same empirical case, the management of irrigated rice terraces by Balinese farmers and villages. In Bali, a fragile system of cooperative management has sustained an equally fragile infrastructure of terraces, tunnels, and aqueducts for many centuries. Nothing like it exists in the tribal societies to the east of Bali. What caused these islands of cooperation to emerge, what sustains them, and why do they sometimes fail? These questions are taken up successively in the fifth, sixth, and seventh chapters. Our goal in these chapters is to draw from decades of research and modeling to distill the lessons learned and clarify what we see as the most useful insights for comparative studies. Lansing's first publications on this topic appeared in the 1970s.

We begin this chapter with an introduction to the origins, ethnography, and ecology of *subaks*, the institutions that, over the course of about a thousand years, terraced the slopes of Bali's volcanoes. The introduction is followed by an exploration of two models that explore the inner workings of the subak system. The first model consists of an abstract game of cooperation involving two players. Like all game-theoretical models, it is designed to capture the essence of strategic decisions by pruning away the details. But unlike other models of cooperation, in which the "environment" consists only of the other players, here it expands to include the natural environment. Consequently, the strategic payoffs for cooperation crucially depend on how the strategies chosen by all the players affect the natural environment. As we will see, the model yields counterintuitive predictions that nicely dovetail with empirical results.

The second model embeds the logic of the first into a simulation of the emergence of local systems of cooperation among 173 subaks that depend on the flow of irrigation water from two Balinese rivers. This model leaves the game-theoretical logic of cooperation untouched, but adds realism by replacing the idealized model of the environment with actual measured flows of rivers, irrigation, rice pests, and harvests. This serves two purposes. First, it enables us to analyze how the environment affects the emergence of varied spatial patterns of cooperation. Second, the predictions of this model can be compared with empirical data. Those results look pretty good, and they look even better in light of an accidental experiment caused by the introduction of Western farming techniques (the Green Revolution), which disrupted the connections between environmental signals and irrigation schedules.

In this first chapter about the subaks, we focus on the interactions between farmers and ecological processes that turn subaks into *coupled human-natural systems*. The next chapter shifts gears to ask, how do higher-level systems of control emerge from the local interactions of groups of subaks? And the third chapter asks two questions: why are some subaks more successful than others, and what determines the transition between successful and less successful regimes? The scale ranges from centuries to days and from river basins to individual farmers.

Terracing volcanoes

We begin with the historical origins of the subak system in Bali. Between the fifth and tenth centuries AD, dozens of little kingdoms came into existence in the islands of Indonesia. Some were focused on trade, like modern Singapore, while others grew up around inland rivers that could be used to grow paddy rice. Most of them survived for only a short time and left few traces, often no more than a few fragmentary inscriptions. The handful of agrarian kingdoms that prospered were situated in the regions that were best suited for rice agriculture. Paddy rice needs an abundant supply of water for flooding the terraces at the beginning of its planting cycle. This requirement was easily met in most of the region. But rice also prefers a reliable dry spell for ripening at the end of the growing cycle and volcanic soil rich in mineral nutrients. Among the large volcanic islands of Indonesia, Java and Bali possessed the best conditions.

The landscape of Bali is dominated by two active volcanoes, with steep slopes reaching almost to the sea. According to Balinese legend, these symmetrical peaks are fragments of the cosmic mountain that were brought to the island by the Hindu gods. Villages and the residences of the rajahs and princes are located on their slopes. There are no irrigation tanks, but four crater lakes at the summits of the volcanoes form natural reservoirs. The flanks of the volcanoes are deeply incised by ravines containing fast-flowing rivers and streams and small diversionary dams, or weirs. The weirs begin near the maximum elevation where rice will grow, and follow at intervals of a few kilometers along each river until they reach the coast. Each weir diverts the flow from a short stretch of river into a small irrigation tunnel no taller than a man and about a meter in width. The tunnels angle sideways and emerge a kilometer or more downslope to flood one or more patches of rice terraces that have been carved into the flanks of the volcanoes. The terracing and tunneling of the volcanoes has gone on for more than a millennium; irrigation tunnel builders are mentioned in ninth-century royal inscriptions.

Unsurprisingly, the earliest irrigation systems were built to take advantage of the most accessible springs and streams, where water flows could be managed with simple technology. Later, as tunnels and canals proliferated on the slopes of the volcanoes, the irrigation systems began to link communities, like melons on a vine. When that occurred, the task of enforcing equitable rules for irrigation schedules would have required cooperation at two levels: among the farmers on each terraced hillside, and between groups of farmers who share the same irrigation system. References to a specialized institution for this purpose begin to appear in Balinese inscriptions in the eleventh century. This institution, called a *subak*, is not identical to the village, which at that time went by the Austronesian name wanua. Instead the membership of a subak consists of all farmers who own land watered by a common source, like a spring or canal.[9] Subaks quickly proliferated in Bali, but never developed in neighboring Java, where irrigation systems were built on gently sloping terrain and remained managed by the villages (wanua).

The immediate goal of the subaks is to manage the flow of water. Rice paddies are artificial ponds that must be kept flooded while the plants are growing, but then dried out for harvest. This imposes strict requirements for the management of irrigation water, and puts downstream communities at the mercy of their neighbors upstream, who by virtue of their location are always in a position to control the flow. Yet the majority of Balinese rice-farming communities are located downstream from others—often at a great distance from their water source—so a workable solution to this problem must have been discovered soon after the farmers began to terrace the volcanoes. Conceivably the solution could have been a top-down system of water management by the rulers. But the sheer number of these small-scale irrigation systems and the need for continuous, intensive management would have made it difficult for the rajahs to control them. Instead, ancient royal inscriptions indicate that the rulers sensibly chose to leave the farmers in control, while taxing the fruits of their labors.

Cooperation between subaks is facilitated by their participation in networks of water temples, which are managed by the subaks. These temples are physically and symbolically linked to the sites where water originates, such as lakes, springs, and the weirs where irrigation systems begin. They are dedicated to deities associated with fertility and growth, notably the Goddess of the Lakes and the Rice Goddess. All of the farmers who benefit

---

[9] Cf. the Baturan inscription (AD 1022, Goris #352), the Er Rara inscription (AD 1072), and the Udayapatya inscription (AD 1181), in M. M. Sukarto K. Atmodjo. 1986. Some short notes on agricultural data from ancient Balinese inscriptions. In Sartono Kartodirdjo, ed. *Papers of the Fourth Indonesian-Dutch History Conference*, Yogyakarta 24–29 July 1983, Vol. 1, *Agrarian History*. Gadjah Mada University Press, pp. 25–62. The Er Rara inscription mentions fields located in at least 27 named hamlets, belonging to 18 different communities, in connection with the kasuwakan (subak) of Rawas. Similarly, Udayapatya refers to 19 kasuwakan. Soekarto locates these kasuwakan in a region stretching from Lake Batur south into Gianyar. Wisseman Christie comments that "the number of kasuwakan involved suggests a local irrigation network of considerable complexity and more than one water source, used by a community spread over a somewhat awkward landscape." W. Christie, Irrigation in Java and Bali before 1500, unpublished ms.

from a particular flow of water share an obligation to provide offerings at the temples where their water originates. For example, if six subaks obtain water from a given weir, all six belong to the congregation of the water temple associated with that weir. Thus, the larger the water source, the larger the congregation of the water temple. The architectural symbolism of the temples identifies each terraced hillock as a miniature replica of the central volcano, which in turn is identified with the mythical mountain celebrated as the *axis mundi* in Hindu and Buddhist mythology. Rituals performed in the temples embellish this cosmic-mountain symbolism by emphasizing the role of the volcanic crater lakes as the symbolic origin of water, with its life-giving and purificatory powers.

But the water temples are more than sites for worshiping the gods. Because the temple networks exactly match the physical system of terraces and irrigation works, they provide a framework for the subaks to coordinate their irrigation schedules. Most subaks have regular monthly meetings, often held in the forecourt of one of their water temples, where they compare information about harvests and pests, agree upon schedules for planting and irrigation, and organize these schedules at the appropriate scales.

Figure 5.1 shows a typical example of a water temple system in the upper reaches of the Petanu River in central Bali. As the figure shows, the Bayad weir provides water for about 100 hectares of rice terraces organized as a single subak. A few kilometers downstream from the Bayad weir is the Manuaba weir, which provides water for 350 hectares of terraces managed by ten subaks. The water temple hierarchy at Bayad consists of a weir-shrine (*Pura Ulun Empelan*) and a "Head of the Rice Fields" temple (*Pura Ulun Swi*) situated above the terraces. The larger Manuaba system also begins with a weir-shrine, but includes two Pura Ulun Swi temples, one for each major block of terraces. The congregations of both Pura Ulun Swi temples also belong to a larger Masceti temple that is symbolically identified with the entire Manuaba irrigation system. Representatives of the ten subaks under the two Pura Ulun Swi temples meet once a year at the Masceti temple to decide on a cropping pattern and irrigation schedule. The degree of nested control apparent in the above description is typical of the overall temple system.

Ecology of the rice terraces

The hundreds of small-scale irrigation systems that capture the flow from streams, rivers, and springs transport dissolved minerals directly to the fields. The volcanic landscape contains significant amounts of phosphorus and potassium that are susceptible to leaching. The deposition of volcanic

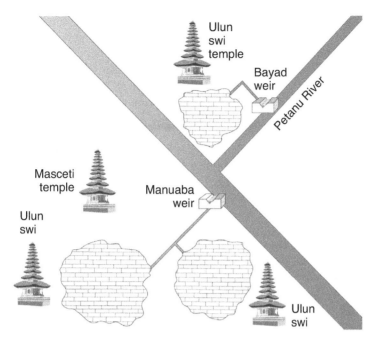

Figure 5.1. An example of the relationship between subaks and water temples on the upper Petanu River. The Bayad weir provides water for about 100 hectares of rice terraces organized as a single subak. A few kilometers downstream from the Bayad weir is the Manuaba weir, which provides water for 350 hectares of terraces managed by ten subaks. The water temple hierarchy at Bayad consists of a weir-shrine (*Pura Ulun Empelan*) and a "Head of the Rice Fields" temple (*Pura Ulun Swi*) situated above the terraces. The larger Manuaba system also begins with a weir-shrine, but includes two Pura Ulun Swi temples, one for each major block of terraces. The congregations of both Pura Ulun Swi temples also belong to a larger Masceti temple that is symbolically identified with the entire Manuaba irrigation system. Representatives of the ten subaks under the two Pura Ulun Swi temples meet once a year at the Masceti temple to decide on a cropping pattern and irrigation schedule. Credit: Yves Descatoire.

ash at frequent intervals, together with weathering caused by wind and rain, gradually but continually replenishes soils with these nutrients. As a result, water used for irrigation provides a direct pathway for fertilization of irrigated rice. Rice paddies have plough pans that trap water and nutrients, creating pond-like aquatic ecosystems where erosion and percolation (and thus loss of nutrients) are minimized. Careful control of the flow of water into the fields creates pulses in several important biochemical cycles necessary for growing rice. Water cycles have a direct influence on soil pH, temperature, nutrient circulation, aerobic conditions, microorganism growth, and weed suppression. In general, irrigation demands are highest at the start of a new planting cycle, since the dry fields must first be saturated with water.

The flooding and draining of blocks of terraces also has important effects on rice pests, including insects, rodents, and bacterial and viral diseases. The problem of pest control is not a recent development: traditional Balinese texts, such as the *Dharma Pamaculan*, refer to rice pests (*hama merana*), and both Balinese and Dutch colonial sources mention devastating plagues of rats in the paddy fields. But if farmers with adjacent fields can synchronize their cropping patterns to create a uniform fallow period over a sufficiently large area, rice pests will be temporarily deprived of their habitat, and their populations can be sharply reduced. Field data indicate that synchronized harvests can keep losses from pests to < 4% compared to losses that can exceed 90% when unsynchronized fallow periods allow pests to migrate from field to field. How large an area must be fallow, and for how long, depends on the characteristics of specific pests.[10] Yet if too many farmers follow an identical cropping pattern in an effort to control pests, then their peak water demands will coincide, and there may not be enough water to sustain flooding over all the fields.

A cooperation game

To gain insight into the trade-offs between water sharing and pest control, Lansing and economist John Miller created a model using game theory.[11] Suppose that there are only two rice farmers, one upstream from the other. The upstream farmer has first claim on any water in the system. To simplify matters, suppose that farmers must choose one of two possible

[10] I. N. Widiarta, Y. Suzuki, H. Sawada, and F. Nakasuji. 1990. Population dynamics of the green leafhopper, *Nephotettix virescens* Distant (Hemiptera: Cicadellidae) in synchronized and staggered transplanting areas of paddy fields in Indonesia. *Researches in Population Ecology* 32:319–28; I. G. N. Aryawan, I. N. Widiarta, Y. Suzuki, and F. Nakasuji. 1993. Life table analysis of the green rice leafhopper, *Nephotettix virescens* Distant (Hemiptera: Cicadellidae). An efficient vector of rice Tungro disease in asynchronous rice fields in Indonesia. *Researches in Population Ecology* 35:31–43; J. Holt and T.C.B. Chancellor. 1996. Simulation modeling of the spread of rice Tungro virus disease: The potential for management by roguing. *Journal of Applied Ecology* 33:927–36; D. R. Latham. 1999. Temporal and spatial patterns of asynchronous rice cropping and their influence on pest and disease occurrence in a Balinese landscape: Developing predictive models of management. MS thesis, University of Michigan at Ann Arbor.

[11] J. S. Lansing and J. H. Miller. 2005. Cooperation games and ecological feedback: Some insights from Bali. *Current Anthropology* 46:328–34.

Table 5.1

Payoffs for the cooperation game. The first value in each box is the payoff for the first farmer, followed by the payoff for the second farmer. So in the upper left box, where both the upstream and downstream farmers choose the A planting schedule, the payoff for the upstream farmer is 1, while the downstream farmer's payoff is $1 - \delta$.

	A_d	B_d
A_u	$1, 1 - \delta$	$1 - \rho, 1 - \rho$
B_u	$1 - \rho, 1 - \rho$	$1, 1 - \delta$

dates on which to plant their crops, A or B. As in the Balinese ecosystem, we assume that the water supply is adequate to accommodate the needs of a single farmer during any given period, but it is insufficient if both decide to plant simultaneously. Let δ ($0 < \delta < 1$) represent the crop loss due to reduced water inputs experienced by the downstream farmer if he plants at the same time as the upstream farmer.

If the farmers choose to plant at different times, both fields will suffer pest damage because pests can migrate between them. Let ρ ($0 < \rho < 1$) represent the crop loss due to pest migration between the fields under these conditions (we assume that there is no damage if the crops are planted simultaneously). Then the payoff matrix of the associated game is given in Table 5.1, where the rows (columns) represent the choices of the upstream and downstream farmer, subscripted by u and d, respectively. Here the payoff is the harvest, normalized to one.

Intuitively, the model's underlying logic is simple. There are two important external forces in the system: water damage (δ) imposed by the upstream farmer on the downstream farmer; and pest damage (ρ) imposed by both farmers on each other by staggered cropping. The upstream farmer is not impacted by water scarcity and thus always has an incentive to minimize pest damage by simultaneous cropping. The downstream farmer faces either water scarcity (under simultaneous cropping) or pest damage (under staggered cropping) and thus will choose the lesser of two evils. If pest losses are low, the downstream farmer wants to stagger cropping due to water considerations, while the upstream farmer wants to plant simultaneously to avoid pest damage. A mixed-strategy ensues. If, however, pest losses are high, both farmers' incentives are to coordinate on one of the two possible simultaneous cropping patterns.[12]

---

[12] The Nash equilibria of this game provides several insights. The game always has a single, mixed-strategy Nash equilibrium in which both players randomize with equal weight over the two starting times. The expected aggregate crop yield from the mixed strategy is $2 - \delta/2 - \rho$. Two pure strategy equilibria (either both planting at time

Nash Equilibrium

The Nash equilibrium is an idea dating from the nineteenth century, but formalized by John Forbes Nash, Jr. in his 1951 paper "Non-cooperative games" (*Annals of Mathematics* 54:286–95).

At a Nash equilibrium, each player chooses an optimal strategy given the strategies chosen by the other players. Thus, at an equilibrium, no player has an incentive to change behavior. At any configuration that is not a Nash equilibrium, at least one player can strictly benefit by changing their behavior. It is possible for a system to have multiple Nash equilibria.

Thus, if pests are bad enough (that is, $\rho \geq \delta$), a coordinated solution emerges with both farmers receiving higher individual crop yields than they would expect under the mixed-strategy outcome. Given that the two resulting pure-strategy equilibria of the coordination game yield identical outcomes, both of which are better than the mixed-strategy outcome, there is an obvious role for an external coordination device (here, the subak meetings) for determining which of the two equilibria strategies to play. It is particularly noteworthy that such an entity does not require any formal enforcement power to remain credible; it is in the individual interests of the farmers to follow whatever edict they collectively choose to impose upon themselves in the water temple (formally, this is known as a coordinated equilibrium).

There is also a range of parameters under which the aggregate yield is likely to improve if more pest damage occurs. This paradoxical result occurs when $\delta > \rho > \delta/2$. In this range of ρ, either of the coordinated outcomes has higher aggregate crop yields than the mixed-strategy outcome, but only the mixed-strategy equilibrium is supported. Under such circumstances, if we increase ρ to ρ' (such that $\rho' > \delta$), the two pure-strategy equilibria become supported and aggregate output can be increased if one of them is adopted by the farmers. When crops are staggered, the aggregate yield falls due to pest damage to both fields, as opposed to

---

A or both planting at time B) arise when $\delta \leq \rho$. Thus, when $\delta \leq \rho$ the game can take the form of a simple coordination game in which the two players would like to plant at the same time. In either of the coordinated equilibria, the aggregate production is equal to $2 - \delta$. Note that the coordinated outcome will yield a greater aggregate harvest than the mixed strategy outcome when $\rho > \delta/2$. This holds since pest damage is borne by both farmers, while water damage only impacts the downstream farmer, thus aggregate yields increase by coordinating when pest damage is at least half as bad as water damage.

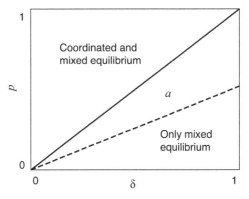

Figure 5.2. Equilibria in the cooperation game. Let δ $(0 < \delta < 1)$ represent the crop loss due to reduced water inputs experienced by the downstream farmer if he plants at the same time as the upstream farmer. If the farmers choose to plant at different times, both fields will suffer pest damage because pests can migrate between them. Let ρ $(0 < \rho < 1)$ represent the crop loss due to pest migration between the fields under these conditions (assume that there is no damage if the crops are planted simultaneously). For all parameter values in the region between the dashed and 45-degree lines, such as point a, aggregate output would be greater at either of the pure-strategy cooperative equilibria even though only the mixed strategy is supported. This leads to a rather counterintuitive implication: for any such point the aggregate crop output could be improved by increasing the damage done by pests (that is, by increasing the value of ρ). Credit: John Miller.

simultaneous cropping in which water damage affects only one field. In this range of pest and water damage, the downstream farmer has no incentive to incorporate the pest damage to the upstream field into his decision calculus, and therefore it is possible for the downstream farmer to prefer staggered cropping even though this lowers aggregate yield. But if pest damage increases, the downstream farmer will eventually prefer the water damage triggered by simultaneous cropping to the pest damage caused by staggered cropping. In this way, more pests can cause both farmers to choose to cooperate in synchronized irrigation, and the aggregate yield from the two fields will rise.

Figure 5.2 summarizes these results. Parameter values below the 45-degree line can only support the mixed-strategy equilibrium, while those above this line can, in addition, support the two pure-strategy equilibria. In terms of aggregate crop output, either of the pure-strategy equilibria (cooperation in synchronizing irrigation) results in greater output than the mixed-strategy equilibrium for all parameters above the dashed line. In particular, note that for all parameter values in the region between the dashed and 45-degree lines, such as point *a*, aggregate output would be greater at either of the pure-strategy cooperative equilibria even though only the mixed strategy is supported. This leads to a rather counterintuitive implication: for any such point, we could potentially improve the aggregate crop output by increasing the damage done by pests (that is, by increasing the value of ρ). By increasing pest damage under such circumstances, we can move the system into a regime where coordination becomes a viable strategy, and since pest damage is fully mitigated under coordination, aggregate crop output increases.

There is another potential path to improve aggregate crop output when the parameters are such that the downstream farmer would prefer not to coordinate. Recall from Figure 5.2 that parameters below the 45-degree line can only support the mixed-strategy uncooperative equilibrium. However, there are circumstances in which the upstream farmer may be willing to pass on some of the water in order to induce the downstream farmer to cooperate. For example, suppose that the crop damage due to water can be shared between the two farmers. For this to occur the upstream farmer takes less than the full amount of water, passing it on so the downstream farmer can experience lower crop losses. It can be shown that there are some damage-sharing arrangements in which both farmers will be willing to coordinate cropping for any parameters in the range between the 45-degree and dashed lines in Figure 5.2. Moreover, as the parameters move from the 45-degree line toward the dashed line, the upstream farmer will be forced to provide a more equal distribution of the loss; that is, the water will need to be more evenly shared between the two farmers to make the arrangement work.

Although the model above is intentionally simplified, it appears to be robust to a variety of changes. For example, the introduction of modern high yielding crops (the Green Revolution) can be modeled by multiplying all of the payoffs by a constant; such transformations have no impact on the analysis. Or instead of simultaneous choices, we could allow one farmer to move first in the game. In the case where the farmers' incentives differ, the outcome of the game would depend on who moved first. If they both want to coordinate, then the first move could serve as a coordination mechanism.

In the model, we also assumed that there were just two players: one upstream and one downstream farmer. But the same dynamics should apply to upstream and downstream subaks, each composed of many individual farmers. This assumption could be violated if, say, individual farmers within a given subak free ride on any group agreements, and thereby destroy the ability of a given subak to act as a unified entity. While more explicit models of subak decision-making are of interest, there are some key factors in Bali that tend to enforce subak cohesion.

In particular, given the proximity and low mobility of individual farmers within a given subak, individuals have very long-term interactions with one another across a variety of social and economic realms, ranging from agriculture to marriage, in an environment in which behavior is easily observed by others. In such a world, the long shadow of the future, multiple ties, and easily available information should tend to promote high levels of cooperation. Indeed, survey evidence presented below suggests that farmers believe that the success of their harvests is closely tied to those of fellow subak members. Moreover, subaks have elaborate codified rules that enforce cooperation within the group once a decision has been taken, punishing those individuals who violate the rules with both informal and formal sanctions. Indeed, it is said that "the voice of the subak is the voice of God."

Testing the game-theoretical model

How well does this simple model capture the actual basis for decision-making by the farmers? In the summer of 1998, we carried out a survey of farmers in ten subaks that belong to the congregation of a regional water temple. In each of the ten subaks, we chose a random sample of fifteen farmers. Of these fifteen, five were selected whose fields were located in the upstream part of their subak; five from the middle of the subak, and the last five from the downstream section of the subak. To test the predictions of the game, we simply asked, "Which problem is worse, damage from pests or irrigation water shortages?" The results, shown in Figure 5.3, show that upstream farmers worried more about pests, while downstream farmers were more concerned with water shortages.

The same dynamics recur at the next level up. Not only individual farmers, but whole subaks had to decide whether or not to cooperate. In our sample, six of the ten subaks were situated in upstream/downstream pairs, where the downstream subak obtained most of its water from its upstream neighbor. Thus it was also possible to compare the aggregate response of all the farmers in each downstream subak to the response of their upstream neighbors.

Here also, the upstream farmers were more concerned with potential damage from pests than from water shortages, and so had a reason to cooperate with downstream neighbors. By adjusting their own irrigation flows to help achieve $\delta < \rho$ for their downstream neighbors, the upstream subaks had the power to promote a solution that was beneficial for everyone. In the real world, losses from pest outbreaks can quickly approach 100% after a few seasons of unsynchronized cropping

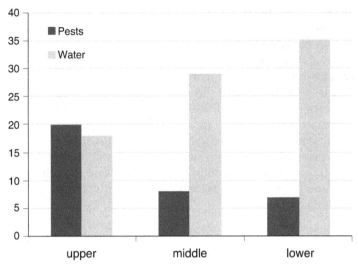

Figure 5.3. Responses of 117 farmers in 10 subaks to the question "which is worse, pest damage or water shortages?" plotted according to the location of their fields within the subak. Pearson $\chi^2 = 14.08$, $P < 0.001$. The y-axis is number of respondents. Credit: J. Stephen Lansing.

schedules. In contrast, reducing irrigation flow by 5% or 10% (or using your own labor to reduce seepage losses in an irrigation system and so improve its efficiency) imposes lesser costs on the farmers, unless water is very scarce indeed. These results are also supported by videotaped records of monthly meetings in which the heads of all ten subaks (plus four others not included in the survey) discussed intersubak affairs. Altogether, the willingness of upstream subaks to synchronize cropping patterns appears to be clearly related to the perceived threat of pest invasions. It is important to note that exactly which subaks synchronize cropping plans with their neighbors varies from year to year. An increased threat of pest damage, such as has occurred recently in several cycles of pest infestations (by brown plant hoppers and rice tungro virus) quickly leads to larger synchronized groups, while a period of light rains encourages greater fragmentation.

We drew two conclusions from this study. First, a very simple model appears to capture the essence of the trade-offs involved in decisions about cooperation among the farmers. Thus, the model suggests an ecological basis for long-term patterns of cooperation. Second, the same dynamics recur at the group level, providing a possible explanation for the aggregation of farmers into subaks and subaks into multisubak groups. In computer simulations of the game with multiple players and realistic ecology, this is exactly what occurs, as shown in the next model.

An agent-based model of the coupled system

Using empirical data on the location, size, and field conditions of 172 subaks in the watershed of the Oos and Petanu rivers in southern Bali in 1987–1988 (Figure 5.4), changes in the flow of irrigation water and the growth of rice and pests were modeled as subaks decided whether to cooperate with their neighbors. In this model, each subak behaves as an adaptive agent that seeks to improve its harvest by imitating the cropping pattern of more successful neighbors.[13]

The model simulates the flow of water from the headwaters of the two rivers to the sea at monthly intervals. The amount of water available for any given subak depends on seasonal patterns of rainfall and ground water flow and the amount of water diverted by upstream subaks for their own needs. As a new year begins, each of the 172 subaks is given a planting schedule that determines which crops it will grow and when they will be planted. As the months go by, water flows, crops grow, and pests migrate across the landscape. When a subak harvests its crop, losses due to water shortages or pests are tabulated. At the end of the year, aggregate harvest yields are calculated for the subaks. Subsequently, each subak checks to see whether any of its closest neighbors got higher yields. If so, the target subak copies the cropping schedule of its (best) neighbor. If none of the neighbors got better yields, the target subak retains its existing schedule.

When all the subaks have made their decisions, the model cycles through another year. These simulations begin with a random distribution of cropping patterns (a typical example is shown in Figure 5.5). After a year, the subaks in the model begin to aggregate into small patches following identical cropping patterns, which helps to reduce pest losses. As time goes on, these patches grow until they become large enough to cause water stress, whereupon patch sizes fluctuate. As they do, harvests also fluctuate but gradually rise. The program continues until all or most subaks have discovered an optimal cropping pattern, meaning that they consistently cannot do better by imitating one of their neighbors.

Online Resource: A Coupled Social-Ecological System Model

The model of coupled social-ecological systems is available to explore in the book's online resources:

https://www.islandsoforder.com/coupled-social-ecological-systems.html

---

[13] J. S. Lansing and J. N. Kremer. 1994. Emergent properties of Balinese water temples. C. Langton, ed. *Artificial Life III*. Addison-Wesley and the Santa Fe Institute Studies in the Sciences of Complexity, Vol. X, pp. 201–25.

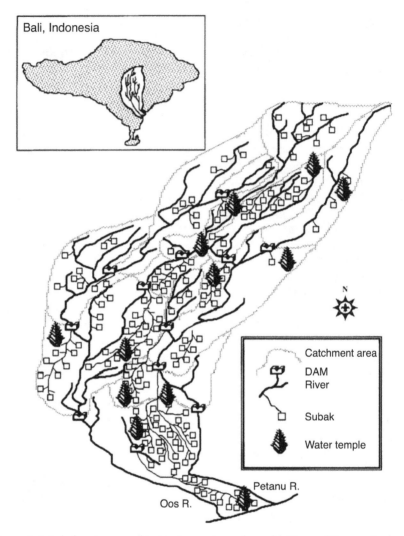

Figure 5.4. Subaks, rivers, and irrigation systems along the Oos and Petanu rivers of southern Bali. Traditionally, each subak is free to choose its own irrigation schedule. By synchronizing irrigation with different-sized clusters of neighboring subaks, crop losses due to pests or water shortages can be avoided. Map is not to scale; subaks are not rectangular; and many more water temples exist than are depicted here. Credit: J. Stephen Lansing.

Figure 5.5. An initial randomized cropping schedule. Symbols indicate cropping schedules for the year; for example, circles might mean "plant rice in January and August," while squares could mean "plant rice in March and October." After a simulated year, the average aggregate harvest (from both crops) is 4.9 tons/hectare. Credit: J. Stephen Lansing.

Experiments with this model indicate that the entire collection of subaks quickly settles down into a stable pattern of synchronized cropping schedules that optimizes the trade-off between pest control and water sharing. The close relationship between this pattern as calculated in the model and the actual pattern of synchronized planting units is clearly apparent (Figure 5.6). In the model, as patterns of coordination resembling the water temple networks emerge, both the mean harvest yield and the highest yield increase, while variance in yield across subaks declines (Figure 5.7). In other words, after just a few years of local experimentation, yields rise for everyone while variation in yields decreases. Subsequent simulations showed that if the environment is perturbed, either by decreasing rainfall or by increasing virulence of pests, a few

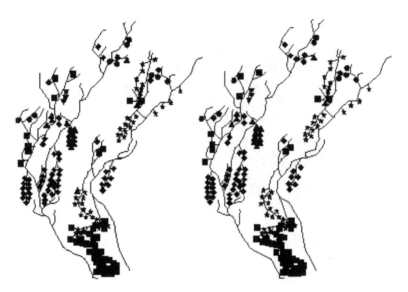

Figure 5.6. Synchronized cropping. Left: After 10 years, synchronized groups of subaks emerge, and average harvests increase to 8.57 tons/hectare. Right: Actual groupings of subaks and their cropping schedules within their water temple networks. Credit: J. Stephen Lansing.

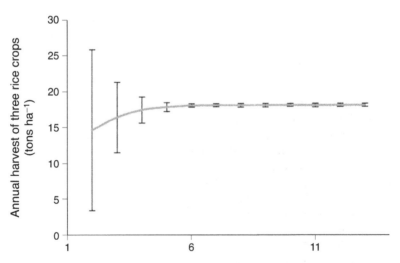

Figure 5.7. Reduction in variance of harvest yields as cooperation spreads in the simulation model of the Oos and Petanu watersheds. Credit: J. Stephen Lansing.

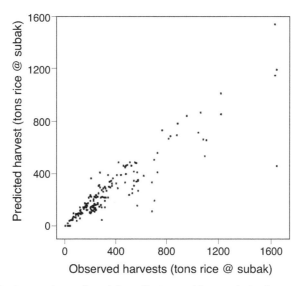

Figure 5.8. Comparison of model predictions with actual rice harvests by subak in 1988–1989 for 72 (of 172) subaks for which data were acquired. The accuracy of the model improved from the first crop of the year ($r = 0.85$) to subsequent harvests ($r = 0.96$), because as time proceeds the accuracy of the pest submodel increased. When the simulation starts, pests are at background levels, and it takes time for the effects of synchronized cropping to affect population levels. Considering the simplicity of the model, yields per hectare were also well correlated with $r = 0.5$. Credit: J. Stephen Lansing.

subaks change their cropping patterns, but within a few years a new equilibrium is achieved.[14]

To validate the model, a field survey was undertaken that obtained two years of data on hydrology, actual planting schedules, and harvest yields from August to December 1988 for 72 of the 172 subaks included in the model, from the headwaters to the sea. The model was calibrated with these data, and the simulated and reported rice harvests were compared for two harvests for 72 subaks in Bali in 1989 (Figure 5.8). The accuracy of the model improved from the first crop of the year ($r = 0.85$) to subsequent harvests ($r = 0.96$), because as time moves on the accuracy of the pest submodel increased (when the simulation starts, pests are at background levels, and it takes time for the effects of synchronized cropping to affect population levels). Considering the simplicity of the model, yields per hectare were also well correlated with $r = 0.5$. To eliminate the

[14] Ibid.

possibility that the model results were simply not responsive to variations in cropping plans, we ran additional simulations in which we disrupted the local coordination implicit in the planting schedules followed by the subaks in 1989. When the cropping patterns were randomized, but the actual crops planted remained the same, the correlation for the second crop in 1989 dropped from 0.50 to just 0.01.[15]

Conclusion

In the 1970s, the Asian Development Bank became involved in an effort to boost rice production in Indonesia. The bank's consultants learned that on Bali, local groups of farmers synchronize their irrigation schedules. In most regions, these schedules produced two rice harvests of native Balinese rice per year. The consultants saw two ways to improve harvests. The first was to encourage the farmers to grow higher-yielding Green Revolution rice varieties, which produce more grain than Balinese rice. The second recommendation took advantage of another feature of the new rice: it grows faster than native Balinese rice. Consequently, the farmers could plant more frequently. The Ministry of Agriculture adopted both recommendations, and competitions were created to reward the farmers who produced the best harvests. Synchronized planting of native Balinese rice was strongly discouraged. Instead, farmers were instructed to plant Green Revolution rice as often as they could. By 1977, 70% of the southern Balinese "rice bowl" was planted with Green Revolution rice, and subaks stopped coordinating their irrigation schedules.

At first, rice harvests improved. But a year or two later, Balinese agricultural and irrigation workers began to report "chaos in water scheduling" and "explosions of pest populations." At the time, planners dismissed these occurrences as coincidence. They urged the farmers to apply higher doses of pesticides, while still competing to grow as much rice per year as possible. This actually intensified both the pest problem[16] and the water shortages.[17] Similar results can be achieved by running the simulation model in reverse: beginning the simulation with the evolved cluster

---

[15] J. S. Lansing and J. N. Kremer. 1993. Emergent properties of Balinese water temples. *American Anthropologist* 95:97–114.

[16] J. S. Lansing. *Priests and Programmers: Technologies of Power in the Engineered Landscape of Bali.* Revised 2nd edition. Princeton University Press, 2007 [1991]; B. Machbub, H. F. Ludwig, and D. Gunaratnam. 1988. Environmental impact from agrochemicals in Bali (Indonesia). *Environmental Monitoring and Assessment* 11:1–23.

[17] L. Horst. 1998. *The Dilemma of Water Division: Considerations and Criteria for Irrigation System Design.* International Irrigation Management Institute, Colombo, Sri Lanka, 1998.

patterns of subaks, and instructing the subaks to plant as often as possible. This quickly leads to the fragmentation of subak clusters, triggering increases in pests and water shortages. It was only when farmers spontaneously returned to synchronized planting schemes that harvests began to recover, a point subsequently acknowledged by the final evaluation team from the Asian Development Bank: "Substitution of the 'high technology and bureaucratic' solution in the event proved counterproductive, and was the major factor behind the yield and cropped area declines experienced between 1982 and 1985.... The cost of the lack of appreciation of the merits of the traditional regime has been high."[18]

[18] Lansing, *Priests and Programmers*, 124–5.

CHAPTER 6

~~~~~~

Adaptive Self-Organized Criticality

> What shall we say of Ops (fortune)? What of Salus (well-being)?
> Of Concordia, Libertas, Victoria? As each of these things has a
> power too great to be controlled without a god, it is the thing
> itself which has received the title of god.[1]
>
> —*Cicero*

Thanks in large part to the models described in the last chapter, by 1990 the functional significance of Bali's water temple system seemed clear to us. The model of the subaks along the Oos and Petanu rivers showed how a water temple network could easily arise and subsequently cope with environmental variation.[2] However, our publications about these results prompted several critiques by other researchers, who questioned the relationship of the temples and subaks to the rulers of Balinese kingdoms. Bali was conquered by the Dutch in a series of colonial wars that began in 1846 and ended in 1908. Were the subaks we observed a twentieth-century innovation? And had we failed to appreciate the role of Balinese rajahs and princes in the construction and management of the rice terraces and irrigation systems?

Responding to these critiques led Lansing to set aside modeling in favor of archival and archaeological research on the history of temples, subaks, irrigations, and kingdoms. This tale is told in two books and several articles, and we will not repeat it here, except to say that in our view there is very little historical evidence that Balinese rulers took an active role in the development of irrigation.[3] But sifting through the historical evidence gradually exposed an apparent gap in our understanding of the

---

[1] Cicero, *Georgics*, 2.61.

[2] This chapter argues that adaptation can trigger self-organized criticality, a nonlinear transition of interest to physicists. For a more technical explanation written for physicists, see J. S. Lansing, S. Thurner, N. N. Chung, et al. 2017. Adaptive self-organization of Bali's ancient rice terraces. *Proceedings of the National Academy of Sciences USA* 114:6504–9.

[3] Lansing. *Priests and Programmers: Technologies of Power in the Engineered Landscape of Bali*, Princeton University Press, 1991; J. S. Lansing. *Perfect Order: Recognizing Complexity in Bali*. Princeton University Press, 2006; J. S. Lansing and T. De Vet. 2012. The functional role of Balinese water temples: A response to critics. *Human Ecology* 40:453–67.

inner workings of the water temples. In general, the relationship of the temples to the irrigation systems is a simple one-to-one map. Each farmer creates a shrine to the agrarian deities near the upstream water inlet to his field. Further upstream, all the farmers who share a common irrigation source belong to water temples that are associated with their subak. The temple of the Goddess of the Lake at Mount Batur is the largest water temple, with the most inclusive congregation consisting of hundreds of subaks. The rituals that take place in the water temples proclaim that this is all one grand system, but the system of control that we modeled begins at the very bottom, in the local interactions of farmers. The rituals proclaim universality, but all the action takes place at the local level. Very likely the claim of universality is simply poetry. But could there be something to it?

While writing his second book about the water temples, Lansing began to experiment with a new model: "universal Bali." The idea was to strip geography from the Oos-Petanu simulation described in the previous chapter in order to focus on the emergence of cooperative groups. In the new model, watersheds were replaced by simpler two-dimensional lattices on which model subaks could interact. Irrigation schedules were symbolized by colors and randomly assigned to subaks on the lattices. Once the simulations begin and the subaks seek to improve their harvests, the colors begin to form mosaic patterns. The idea of using colors to represent irrigation schedules was suggested by aerial photographs of the rice terraces. Viewed from above, a changing mosaic of colors is visible: green when the rice is young, yellow as it nears harvest, silver when the paddies are flooded, and brown when they are drained. By assigning colors to the irrigation schedules of the subaks in the model, it soon became apparent that the mosaics on the lattices follow predictable patterns as they evolve: from a chaos of randomly colored patches to distinctive mosaic patterns. Could these patterns be meaningful?

Mosaics and power laws

The mosaic patterns that appear in the lattice model can be compared with the mosaics visible in aerial imagery of the rice terraces by counting the distribution of different-sized colored patches. Doing so, we saw that when the decision rules in the model strike an equal balance between water sharing and pest control, the lattice patterns in both mosaics (model lattices and observed aerial images) converge. Intriguingly, when this occurs, the distribution of colored patches in the lattices takes the form of a power law: a unique pattern with many small patches and progressively fewer large ones. The appearance of a power law distribution hinted that there might indeed be an underlying principle of universality underlying the evolution of this system. In physics, the notion of universality originated in the study of phase transitions in statistical mechanics. A phase transition occurs when a material changes its dynamical properties: water, as it boils and turns into vapor; or a magnet, which when heated loses its magnetism. Universality classes define systems with identical macro-scale properties that are independent of the small-scale details of the system.

Still, power laws occur for many reasons (not only phase transitions) and are observed in many natural systems. But in both the lattices and the aerial images of the rice terraces, a second power law relationship emerged in conjunction with the first. The colored patches are created by the farmer's decisions about when to synchronize irrigation schedules with their neighbors: each patch displays the outcome of these choices. Simulations showed that when the frequency distribution of patches align in a power law, the correlation distance between patches also follows a specific power law (Figure 6.1). This implies that the scale of synchronized management of water by the farmers has expanded from local groups to the entire system. In the model, harvests reach a maximum at precisely this point, and the simulation stops evolving. As we shall see, this mosaic configuration also matches the distribution of irrigation schedules visible in the aerial imagery. And it happens to coincide with what physicists call the "critical point," equivalent to the phase transition that occurs when water starts to boil or a metal becomes magnetic.

In the physical world, the transition to the critical point is always caused by an external force, like temperature. But change in the lattice model is endogenous, driven by the efforts of the model subaks to improve their harvests. In no known case does it occur as a consequence of an adaptive process. Why, then, is there a transition at the critical point in the model? Why does it occur at the point where harvests are maximized? And most intriguingly, why do the irrigation schedules in the aerial imagery match the distributions in the model at the critical point?

Universal Bali: A lattice model

In the previous chapter, we explored two models: a coordination game between upstream and downstream farmers, and a simulation of irrigation and rice harvests along two Balinese rivers. Here we adopt a slightly different strategy: we embed the logic of the two-player game on a uniform lattice. This allows us to explore the full range of spatial configurations that the game can generate. Recall that in the game, the upstream farmers are allowed to have first claim on any water in the system, and can

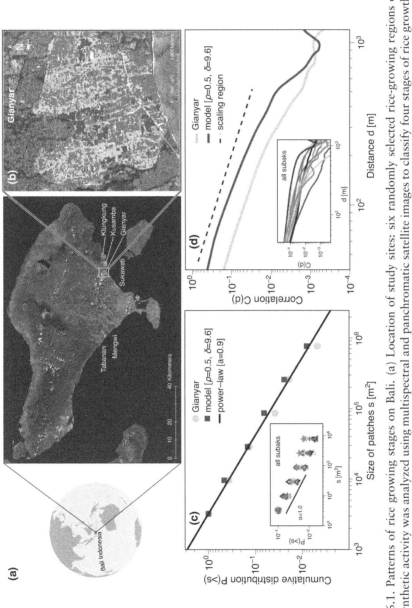

Figure 6.1. Patterns of rice growing stages on Bali. (a) Location of study sites; six randomly selected rice-growing regions of Bali. Photosynthetic activity was analyzed using multispectral and panchromatic satellite images to classify four stages of rice growth in the terraces, which appear as differently colored patches. (b) Image analysis of rice growth, indicating synchronized irrigation schedules in the region of Gianyar. The four shades of the patches indicate the four stages: growing rice, harvest, flooded and drained. (c) Cumulative distribution of the patch sizes for Gianyar (circles) and for model results (squares). Inset: all 13 observations at the six regions, indicating power law behavior, with an exponent around $\alpha \approx 1$. (d) Correlation functions $C(d)$ of the image (planting regions only) as a function of distance for Gianyar and the model. The slow decay (power law) indicates long-range correlations, or *system-wide connectivity* of patches. Inset: all 13 observations. Credit: Authors.

impose water stresses on their downstream neighbors.[4] But harvests are also affected by rice pests. If the farmers plant at different times, they will harvest at different times, and this provides an opportunity for rice pests to migrate between the fields. If the upstream farmer is not very worried about damage from pests, he will have little incentive to synchronize his irrigation schedule with the downstream farmer. This results in a mixed strategy. But both farmers will obtain better harvests by cooperating in a single irrigation strategy. This holds because pest damage is borne by both farmers, while water damage impacts only the downstream farmer. In this case, it is in the individual interest of both farmers to cooperate.

Thus, in the two-player game, whether cooperation emerges depends on the trade-off between pest damage and water shortages, both of which are fixed and known to the players in advance. In reality, for any farmer, pest damages depend on both the intrinsic capacity of endemic pests to cause damage and on whether neighboring farmers choose to control the pests by synchronizing irrigation. Similarly, water shortages depend on both the inflow of irrigation water into the subak and on the scale at which groups of farmers synchronize irrigation. Consequently, the pest-water trade-off for each farmer varies depending on where his farm is located and the outcome of the irrigation schedules chosen by his neighbors. Whether both farmers choose to cooperate (synchronize irrigation) depends on the extent of these stresses.

To explore how patterns of irrigation scheduling emerge from this mutual dependence, an adaptive version of this game was created in which farms are embedded on the sites of an $L \times L$ lattice, with dimension $L = 100$. Parameters ρ and δ specify the relative weights of pest and water stress, respectively, for the entire lattice and are set in advance. The lattice represents a rice growing region such as shown in Figure 6.1b.

This model proceeds through a process of trial-and-error adaptation. Losses from water stress are calculated based on the distribution of irrigation schedules for the entire lattice: the fewer the farmers following a given schedule, the more water they have to share. But this reward for asynchronous irrigation is balanced by the need to reduce losses from pests, which depends on the fraction of neighboring farmers f_p within a given radius r that synchronize their irrigation schedules. When pest damage is at least half as bad as water damage, does cooperation spread and do aggregate harvest yields increase?

The model is initialized with random irrigation patterns for all sites at time $t = 0$, when every farmer i chooses one of four possible irrigation schedules C_i with probability ¼. At the end of a time step (representing

---

[4] Lansing, *Priests and Programmers.*

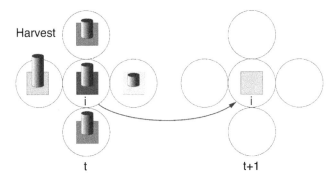

Figure 6.2. Update rule for farmer i. Shades denote different irrigation schedules. For example, one might signify planting in January and another in March. At time $t+1$, farmer i compares his harvest at time t with those of his four closest neighbors. Because the left-hand schedule produced the best harvests, he adopts it for the next cycle. This update corresponds to step 3 in the model. Credit: Authors.

one simulated irrigation cycle), each farmer compares his harvest with those of his closest neighbors, and uses this information to choose his irrigation schedule for the next cycle (Figure 6.2). Because the farmers do not know ρ and δ in advance, they must guess. Anticipating future pest outbreaks or water shortages is challenging, and the actual decision-making process in subaks typically involves lengthy discussions.[5] Irrigation flows along the tiny canals that connect adjacent fields are also complex, involving bargains similar to the game described above. We do not attempt to replicate this level of complexity in the model. Instead we implement very simple strategies to discover whether they are sufficient to enable successful adaptation (Figure 6.2). Once the decision rule and the background pest and water levels are determined, the model proceeds in the following steps:

1. Assume we are at the beginning of time step $t+1$. Calculate the rice harvest for each individual farmer i by debiting his losses from pest damage and water stress, according to $H_i(t+1) = H_0 - \frac{\rho}{0.1 + f_p^i(t)} - \delta f_w^i(t)$, where H_0 is a constant representing the initial harvest before loss. Here, $f_p^i(t)$ denotes the fraction of neighbors of farmer i within a radius r who share the same cropping pattern as i that reduces local pest damage at the previous time step t, and $f_w^i(t)$ is the fraction of *all* lattice sites that have the same cropping pattern as i. The

constant 0.1 in the formula is to ensure that H_i is positive. The parameters ρ and δ specify the relative weights of the pest loss and water stress, respectively. We set $H_0 = 5$ and $r = 2$ (lattice units) for all simulations.

2. Pick one specific farmer i randomly.
3. Farmer i compares his harvest $H_i(t + 1)$ with the harvests of his four nearest neighbors and copies the irrigation schedule of one or more neighbors according to the decision rule (see Figure 6.2). In the simplest case, it is the neighbor who had the best harvest in the previous irrigation cycle j: $C_i(t + 1) = C_j(t)$ (see Figure 6.3).
4. Pick next farmer until all are updated (synchronously).
5. For a small fraction of lattice sites, the irrigation schedules are updated randomly, to simulate empirically observed nonconformity.
6. Move to the next time step, and repeat until harvests converge to a maximum.

Online Resource: The Cooperation Model

The model of cooperation is available to explore in the book's online resources:

https://www.islandsoforder.com/cooperation.html

Results of the lattice model

The model evolves through a process of trial-and-error adaptation by the farmers. At first, in the initial random state ($t = 0$), the correlation between farms is close to zero (Figure 6.3a). What happens next depends on the ecological parameters of pests ρ and water stress δ, and on the decision rule followed by the farmers. There are three trivial attractors (*phases*):

- If water stress is negligible ($\delta \ll 1$), eventually all farms adopt the same irrigation schedule to control pests, resulting in a single uniform patch that spans the entire lattice.
- If $\delta > 20\rho$, water stress dominates and many small patches appear. This increases the variance of irrigation schedules, reducing water stress, but allowing pests to migrate between adjacent patches.
- For $\delta < 20\rho$, after a very long transient phase (thousands of cycles), a *quadrant state* is reached that separates the lattice into four quadrants, each with a different irrigation schedule.

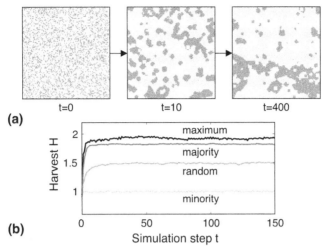

(a)

(b)

Figure 6.3. Evolution of irrigation schedules. (a) Irrigation schedules improve from an initial random configuration at $t = 0$ to $t = 10$, whereupon patch sizes become power law distributed. At $t = 400$, the irrigation patterns change very little and approach a long-lived steady state distribution. (b) Effect of decision rules on harvests. For the "maximum" rule (Step 3, in which farmers choose the best harvest in their neighborhood), average harvests rapidly increase as patch distributions shift to the power law distribution (maximum line). A similar rapid increase occurs for the "majority" update strategy, in which farmers copy the schedule of the majority. Copying a random neighbor's irrigation schedule ("random") leads to inferior harvests. Extending this logic, when farmers update according to the minority of their neighbors, harvests do not improve. The maximum possible harvest is $H = H_0 = 5$ in the absence of pest or water stress. In the simulation shown, both pest and water stresses are strongly present, $\rho = 0.5$ and $\delta = 9.6$. Credit: Authors.

The fourth attractor, which is nontrivial, emerges at the phase transition, exactly at the boundary where the water and pest stress phases equalize. Correlation lengths increase as the cycles of planting and harvest progress, and farms coalesce into small, irregularly sized patches with identical irrigation schedules. Patches form very quickly, as seen in Figure 6.3a, and soon become large enough to dramatically reduce pest damage. Uniformly short correlation distances indicate that the patches are functionally independent: each patch discovers its own solution to the pest-water trade-off. Rice harvests improve rapidly within the first timesteps, and correlations between farms increase. But there is still some variation in harvests, so farms on the borders of the patches continue to experiment with different irrigation schedules. Adaptation ceases when no farm can improve its harvest by changing its irrigation schedule. The geographic scale at which the pest–water trade-off is solved shifts from many small independent patches (small correlation length) to the entire lattice by $t = 10$, equivalent to just five years of double cropping. Subsequently there is little change: at $t = 400$, the situation is very similar to $t = 10$. In Figure 6.3b we study the average harvest $H = \frac{1}{L^2} \sum_{i=1}^{L^2} H_i$ as a function of simulation timesteps (maximum harvest yield strategy). The maximum of H is reached very quickly.

In summary, cooperation quickly spans the entire lattice. Harvests tend to increase and equalize, approaching Pareto optimality at the phase transition where both the frequency distribution of synchronized irrigation patches and the correlations between them become power laws. In the phase diagram for the lattice model, this occurs in a narrow region at the boundary between the regions dominated by pests and water (Figure 6.4).

Pareto Optimality

Pareto optimality (also called *Pareto efficiency*) is a concept proposed by Italian economist Vilfredo Pareto in his 1906 work *Manuale di economia politica* (Società Editrice Libraria). An allocation of resources is Pareto optimal (also called *Pareto efficient*) if any other allocation would make at least one individual strictly worse off. In an economy, Pareto optimality implies output maximization, but the implication does not hold in the other direction. If the output is misallocated, two people could trade and both be better off. Pareto optimality is a widely used concept in economics and game theory. It indicates a point—more often obtained theoretically than in practice—where the overall fitness of the system is maximized.

Comparison with satellite imagery

Patch distributions were analyzed in six rice-growing regions randomly selected on the basis of absence of cloud cover in the images (Figure 6.1a). Figure 6.1b shows one of these regions, Gianyar, on a particular observation day. Four different phases of rice growth corresponding to the irrigation schedules are clearly visible in the multispectral and panchromatic satellite images: growing rice, harvest, flooded, and drained. Image analysis is based on measuring photosynthetic activity. Figure 6.1c shows the cumulative distribution function of the patch sizes s, as they are found in panel b. It shows a clear power law distribution $P(> s) \propto s^{-\alpha}$ with a tail exponent of $\alpha = 0.93 \pm \text{SE } 0.07$. The patch size distributions for all other

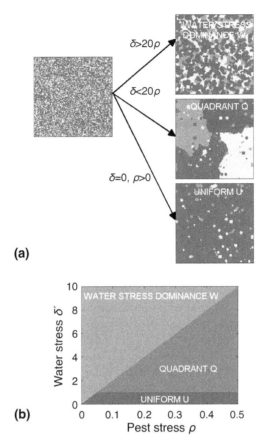

(a)

(b)

Figure 6.4. Phase diagram of the model. (a) Depending on the parameter region in ρ and δ, there exist three separate phases: The "patchy" phase, which is the water stress dominated phase, W. A "quadrant" phase, Q, is found if pest stresses dominate. For the case of no water stress $\delta = 0$, there exists a phase that has a single cropping pattern, the "uniform" phase U. The shape of the phase diagram (b) can be understood by noticing that the relevant parameter in the model is the fraction of the stress factors, δ/ρ. Noise block size is $N = 4$, $t = 4,000$. Credit: Authors.

regions at all observation times are shown in the inset; corresponding exponents are fitted from the data with a standard maximum likelihood estimator and are listed in Table 6.1.

The cumulative patch size distribution is visible in the power law (Figure 6.1c). The model results (squares) for the phase transition (when $\delta/\rho \approx 20$ at $t = 400$) closely matches the empirical data (circles), and would be very similar at $t = 10$. Similar agreement occurs in the

Table 6.1
Power law exponent estimates α (standard errors) and correlation lengths ε (in meters) for 13 measurements (observations) in six rice-growing regions of Bali.

Study Site	2002 α	ε	2013 α	ε	2014 α	ε	2015 α	ε
Gianyar			1.07 (0.08)	697	0.93 (0.07)	373	1.13 (0.07)	261
Klungkung			0.97 (0.07)	472	0.88 (0.07)	511	0.85 (0.07)	336
Kusamba			1.19 (0.09)	762	1.14 (0.09)	552	1.18 (0.14)	607
Mengwi							1.13 (0.08)	579
Sukawati					0.76 (0.09)	1,413		
Tabanan	0.95 (0.09)	507	1.08 (0.17)	566				

correlation function $C(d)$. For the appropriately scaled model results (to match the length scales in the satellite images and the model dimension), we find very similar functional dependence of the correlation function in Figure 6.1d. Both data and model show an approximate power law decay in the correlation function.

Correlation functions $C(d)$ provide a second measure of the scale of cooperation among farmers. In Figure 6.1d for Gianyar, we see that correlation functions decay slowly with distance: the closer two patches are, the more likely they are to follow the same irrigation schedule, indicating that all patches are linked. Correlation functions decline as a power law. Thus, the state of each patch affects all the others, and the Gianyar rice terraces form an integrated (globally coupled) system. The inset shows that this is true for all regions and observations. To quantify the typical correlation length, it is defined as the variance of the correlation function. For Gianyar, the correlation length turns out to be $\varepsilon = 373$ m, spanning all patches. The results for the other regions are found in Table 6.1.

Why power laws?

We suggest that the dynamics captured in the lattice model described above show that self-organized criticality (SOC) can emerge from an adaptive process. The evidence that this tells us something about the Balinese subak system is based on the remarkable similarity of the distributions of patch sizes and correlation distances in the satellite imagery and the model. However, power law distributions can occur for many reasons. For example, they are often found in vegetation patches in dryland ecosystems under stress.[6] But vegetation patches in natural ecosystems are functionally similar, differing only in size. For the vegetation patches that make up the mosaics of the rice terraces, size matters, but so does the age of the rice crop in each patch, which in turn depends on the irrigation schedules selected by the farmers. Any explanation for the observed power law distribution of patches in the rice terraces needs to account for this functional coupling of irrigation schedules and ecosystem dynamics. This adaptive self-organized criticality model tests the hypothesis that the observed mosaic patterns can arise from the farmers' efforts to optimize the pest-water trade-off. The model shows that if the adaptive dynamics are driven by the pest-water trade-off, there exist critical points where the power law distribution is the attractor. Because approximate Pareto optimality emerges at this point, where the pest-water trade-off is optimized at all scales, the model also suggests an explanation for the historical persistence of this attractor. For these reasons, we suggest that the emergence of power law mosaics is not a purely biological phenomenon, but the outcome of ongoing coupled human-natural dynamical interactions. Two further assumptions of the model can be evaluated with historical data.

First, the model assumes that subaks actively cooperate to minimize losses due to pests and water shortages by synchronizing their irrigation schedules. This assumption can be evaluated in light of historical evidence. From the ninth to the fourteenth centuries AD, numerous royal inscriptions encouraged villagers to construct irrigation systems and left water management in their hands.[7] Because of Bali's steep volcanic topography, "the spatial distribution of Balinese irrigation canals, which by their nature cross community boundaries, made it impossible for irrigation to be handled at a purely community level."[8] Later on, both Balinese and European manuscripts describe cooperative management by

[6] R. V. Solé and S. C. Manrubia. 1995. Self-similarity in rain forests: Evidence for a critical state. *Physical Review E* 51:6250–3; M. Pascual and F. Guichard. 2005. Criticality and disturbance in spatial ecological systems. *Trends in Ecology & Evolution* 20:88–95; S. Kéfi, M. Rietkerk, M. Roy, et al. 2011. Robust scaling in ecosystems and the meltdown of patch size distributions before extinction. *Ecology Letters* 14:29–35; S. Kéfi, M. Holmgren, and M. Scheffer. 2016. When can positive interactions cause alternative stable states in ecosystems? *Functional Ecology* 30:88–97.

[7] V. I. Scarborough, J. W. Schoenfelder, and J. S. Lansing. 1999. Early statecraft on Bali: The water temple complex and the decentralization of the political economy. *Research in Economic Anthropology* 20:299–330.

[8] J. W. Christie. 2007. *A World of Water: Rain, Rivers and Seas in Southeast Asian Histories*. KITLV Press, 2007, pp. 252–3.

the subaks. Soon after the final conquest of Bali by the Dutch in 1908, the colonial irrigation engineer tasked with surveying Balinese irrigation wrote: "if due to lack of water not all areas can get water, then they create a turn-taking which is decided upon during the monthly meetings."[9]

Second, the model predicts that rice yields will be optimized by irrigation schedules that balance the pest-water trade-off for multisubak groups. This prediction was inadvertently tested by the introduction of Green Revolution agriculture to Bali in the 1970s. The previous chapter detailed how disastrous this natural experiment was.

Conclusion

Why was the functional significance of multisubak cooperation not apparent to the Green Revolution planners? The process of adaptive self-organized criticality that we observe in the lattice model suggests a possible explanation. Power law distributions of dryland vegetation are comparatively obvious because the patches differ only in size. But adaptive management by the subaks creates differentiated patches of varying size. The distinction is significant, not only because similar versus differentiated patches can occur for different reasons, but also because it is harder for observers to detect the connectivity of differentiated patches. Perhaps partly for this reason, until now, theoretical models of coupled human-natural systems like rice terraces have not anticipated or accounted for the emergence of global-scale connectivity, focusing instead on local interactions. The model also suggests an explanation for the widespread occurrence of fragile kilometers-long irrigation systems linking multiple subaks in the mountains of Bali. If management by the subaks were purely local, leaving downstream subaks at the mercy of their upstream neighbors, these irrigation works would be pointless, and the total area of terraced fields on the island could never have reached its historic extent.

In retrospect, it is perhaps not surprising that the concept of self-organized criticality is relevant to the emergence of cooperation in human interactions with ecosystem processes. Models of self-organized criticality were developed to understand when small-scale local interactions can transition to integrated global connectivity, popularized by the compelling sand-pile example described in chapter 1.[10] These models often behave as if operating exactly at a phase transition. There the systems

---

[9] E. J. van Naerssen. 1918. Irrigatie en waterverdeling volgens de opvatting der Baliërs. *Adatrechtbundel* 15:27–39.

[10] P. Bak. *How Nature Works: The Science of Self-Organized Criticality.* Copernicus, 1996.

become "critical," which means that correlations become long-range and effectively span the entire system, even though interactions only happen at the local nearest-neighbor scale.

In the lattice model, realistic configurations of patches appear after just a few simulation steps. At the same time, harvests approach Pareto optimality—if any farmer changes his irrigation pattern, his rice harvests or those of other farmers will decline. The total harvest of all farms is also maximized. The subak model does not evolve to full alignment of behavior (except when water stress $\delta = 0$), which would minimize pest losses but maximize water stress. Instead, at the critical point the adaptive update process of farmers continues to a point where correlations span the entire system. For this reason, we call the model dynamics *adaptive self-organized criticality*.

For both upstream and downstream farmers, whether or not cooperation is their best strategy depends on the balance between pest and water stress. In the lattice model the adaptive selection of irrigation schedules by individual farmers equalizes water sharing at the phase transition. The game and the lattice model are not directly comparable, because pest and water stresses are calculated differently. But they offer complementary insights: the game captures the logic of the pest-water trade-off, while the lattice shows how cooperation can spread in a coupled system, where farmers adapt to the pest and water stresses triggered by their own decisions. As the phase portrait shows, across a wide range of parameter values, local adaptation will reduce both pest and water stress, initially in local neighborhoods. At the phase transition, these stresses balance each other while harvests are optimized (Figure 6.5). The equalization of water sharing in the lattice model is not assumed from the start, but emerges at the phase transition within a certain parameter range.[11] The resulting mosaics of correlated irrigation schedules neatly mirror the satellite imagery.

How well does the simple decision rule in the lattice model (imitate your most successful neighbor) capture the process by which farmers and subaks adjust their irrigation schedules? In reality, the farmer's decisions reflect the imperatives of the terraced landscape, where fields must be kept flat and protected by bunds to turn them into shallow ponds. The average farm is about 0.3 hectares and consists of many small adjacent ponds. Peak irrigation demand occurs at the beginning of each planting cycle, to create the ponds. Afterwards, the tiny irrigation channels that

---

[11] Irrigation schedules are randomly allocated to farms at the start of the simulation. Subsequently they fluctuate during trial-and-error local adaptations until the model reaches its attractor. They nearly equalize only at the critical transition and quadrant states.

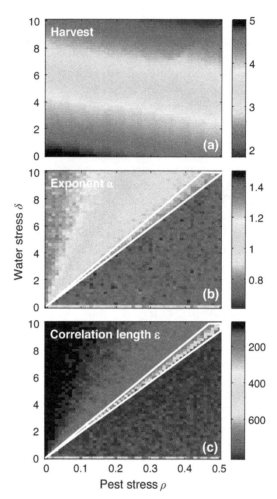

Figure 6.5. Effects of pest and water stress: model results as a function of parameters pests ρ and water shortages δ. (A) Average harvests. The maximum possible harvest H_0 occurs when $\rho = \delta = 0$. (B) power law exponent α of the cumulative patch size distribution. The parameter region that matches the observed slopes from the satellite imagery (Table 6.1) is indicated by the white line. (C) Correlation length ε. The parameter region that matches the observed slopes from the satellite imagery (Table 6.1) are found where $\delta/\rho \sim 20$, which is indicated with the white line. Further computations show the same critical behavior at $\delta/\rho \approx 14$ when $m = 0.2$, or at $\delta/\rho \approx 24$ for $m = 0.05$. The constant m bounds the maximum pest stress at $f_\rho = 0$. Thus, the emergence of critical behavior does not depend simply on ρ and δ, but also on the constant m in the denominator of pest stress. In conclusion, taking results from exponents and correlation lengths, the parameter region that is compatible with observations is $\delta/\rho \approx 20$. Simulations were performed with $L = 100$, $r = 2$, $f = 0.05$, and $N = 4$. Credit: Authors.

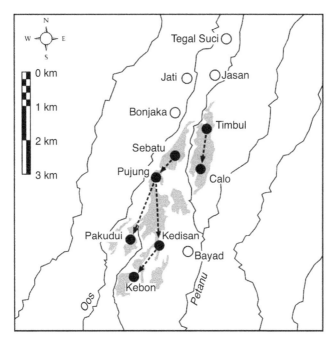

Figure 6.6. Locations and water relationships of subaks on the Petanu River that coordinate irrigation schedules at the water temple Masceti Pamos Apuh. Measured flows are shown in Table 6.2. All rivers shown flow south. A royal inscription provisionally dated to the twelfth century mentions contributions made by the irrigation leaders of several of these subak (Sebatu, Kedisan) to ceremonies in the village where their water originates. Credit: John Schoenfelder.

Table 6.2

Average measured flow volumes at the intake to primary canals (liters per second) compared with water rights based on proportional shares (*tektek*) in July 1997 and 1998 (height of the dry season) among the subaks of the Masceti Pamos Apuh. The correlation coefficient $r = 0.997$. The average flow into each subak was 2.8 liters per second per hectare with a standard deviation of 0.9 liters/sec/ha.

Subak	Flow (L/Sec)	tektek
Jati	71	1.5
Bonjaka	102	2.5
Bayad	198	7.0
Tegal Suci	190	7.5
Pujung	198	8.0
Kedisan	214	8.0
Jasan and Sebatu	386	16.0
Timbul and Calo	460	21.0

connect the ponds require continuous monitoring. Farmers often borrow water from their upstream neighbors; a debt that can be repaid later on by temporarily blocking the flow to their own fields. A farmer who cannot borrow water from one upstream neighbor can try to borrow from others whose fields are either adjacent to the first upstream neighbor or further upstream. For these reasons, decisions about water sharing, irrigation schedules, and the need for pest control begin with conversations among small groups of neighboring farmers. Importantly, this is true for upstream farmers as well as those whose fields are located downstream. Subak meetings provide a venue to reach a consensus. An analogous process occurs in the lattice model, as neighbors create synchronized patches that eventually become correlated, equalizing water sharing.

If several subaks share water resources, their elected leaders meet to negotiate irrigation schedules. Although this higher-level coordination between subaks is not explicitly included in the model, the decision-making process is the same: a trial-and-error adaptation to reduce pest and water stress. These meetings take place in regional water temples and make use of a sophisticated permutational calendar to plan and implement staggered irrigation schedules.[12] These cultural innovations undoubtedly facilitate adaptation to changing pest-water dynamics. But the model does not require calendars or water temples; instead it helps to clarify their functional significance for sustaining approximate Pareto optimality. Our model shows that the simple pest-water trade-off triggers continuous transitions that turn adaptive agents on a two-dimensional lattice into a coevolving system capable of solving the pest-water trade-off by means of local decision-making. Unlike Lance Gunderson and C. S. Holling's well-known model of adaptive cycles,[13] here increasing connectivity does not cause collapse, but stabilizes at a scale-free distribution of functionally varied patches. This is quite a general result that may be common in coupled human-natural systems. In any anthropogenic landscape, correlations between patches will provide some information about the scale of human management. If Bali's subaks are not unique and adaptive self-organized criticality occurs in the management of the commons elsewhere, it should be readily detectable from correlated patch distributions.

We conclude with the question of whether these results are likely to be unusual, perhaps even unique to Bali. The scope of the model is limited by the physical geography of Bali. The four crater lakes store rainfall that feeds the groundwater system, but they have no river outlets. On

---

[12] Lansing and De Vet, Balinese water temples, 453–67.

[13] L. H. Gunderson and C. S. Holling. *Panarchy: Understanding Transformations in Human and Natural Systems.* Island Press, 2001.

the steep porous volcanic slopes, rivers recharge very quickly. Irrigation systems consist of closely spaced weirs and springs that provide water for one or more subaks. These local irrigation systems are functionally independent: although they remove most or all of the flow, a kilometer or two downstream it will be replenished from groundwater flows. Our model captures the adaptive process at this scale, where local groups of farmers meet face to face to solve the pest-water trade-off. The concept of emergent global-scale connectivity in our model, which we borrow from physics, does not refer to all the subaks on a river, but to these smaller functionally independent groups of subaks, such as those shown in Figures 6.1b and 6.6. This contrasts with a typical desert river, where the effects of upstream irrigation may be felt far downstream.

~~~~~~

CHAPTER 7

Transition Paths

> The voice of the subak is the voice of God.
> —*A traditional Balinese saying*

Introduction

In the long run, there are strong incentives for Balinese subaks to sustain cooperative management of their rice terraces. But in the short run, cooperation sometimes falters, usually for brief periods but occasionally permanently. How predictable are these failures, both in terms of their causes and their frequency? Our attempts to answer this question led us to some new paths in the adaptive landscape. As we saw in the last chapter, rice paddies are shallow artificial ponds that must be flooded and drained in synchronous patches to deliver nutrients and promote plant growth, while also controlling weeds and rice pests. Achieving these results requires carefully timed collective action by groups of farmers. Environmental historians have long argued that under such circumstances "the reciprocal influences of a changing nature and a changing society" will cause landscapes and institutions to coevolve.[1] But this is not always a simple process, a point that was recently brought home by the struggles of farmers in the Netherlands trying to manage their lakes. For decades, excess fertilizer flowed into the lakes, triggering algae blooms and eutrophication. But as the farmers discovered, simply reducing the amount of fertilizer entering the lakes was not enough to restore them to clarity. It turned out that alternate stable states existed, one turbid and the other clear. In ecology, such alternate stable states are known as *regimes*. But once the existence of these alternate regimes was recognized, a simple intervention was sufficient to restore the lakes to health: temporarily removing the fish allowed sediment to settle and zooplankton populations to increase, whereupon water clarity could be improved by reducing the amount of fertilizer flowing into the lakes.[2]

---

[1] R. White. 1985. American environmental history: The development of a new field. *Pacific Historical Review* 54:297–335.

[2] This discovery was recognized with the awarding of the Spinoza Prize in 2009 to mathematical ecologist Marten Scheffer. See J. L. Attayde, E. H. Van Nes,

However, the management of rice paddies is considerably more complicated. From one year to the next, Balinese farmers must adjust to the changing environmental and social conditions on which their livelihoods depend. This requires ongoing collective action, which typically includes a heavy burden of ritual obligations, as well as agricultural labor. Too little investment of effort in the subak risks triggering crop losses, angry neighbors, or indeed the wrath of the gods, but too much investment might incur the wrath of one's family. Lansing's ethnographic observations suggested that subaks vary in their ability to increase or decrease these investments, as conditions require.[3] When this capacity declines, the subak becomes vulnerable. But failures in cooperation may be temporary; crop losses may prompt a return to high investments in the subak. Thus, while low investment could mean that a subak is close to collapse, it could also mean that the farmers are enjoying a period of low stress, which may or may not be transient.

From a comparative perspective, subaks may be particularly well suited for the analysis of coupled social-ecological interactions for three reasons. First, they are functional institutions: the prosperity of traditional Balinese rice-growing villages mostly depends on their efficacy. Second, the time lag between failures in cooperation within the subak and perceptible consequences in the paddy field is short due to the small scale of subak irrigation, the fragility of terraced fields, and the potential for crop losses from pests or water shortages. Hence there is a strong potential for social learning and adaptation. Third, there is large variation in relevant variables such as the age of subaks, their demographic composition, and local environmental conditions: for centuries, subaks have evolved independently in neighboring catchments.

So what causes failures in cooperation to occur? Based on many years of ethnographic observations, we predicted that—independent of environmental conditions—pro-social behavior might be more common in subaks where nearly all families share a long history of managing their rice terraces, bringing with it a heightened awareness of the consequences of failure. If so, older and more cohesive subaks might be less prone to failures than newer ones. But can the age of subaks be determined?

A.I.L. Araujo, et al. 2010. Omnivory by planktivores stabilizes plankton dynamics, but may either promote or reduce algal biomass. *Ecosystems* 13:410–20.

[3] J. S. Lansing. *Perfect Order: Recognizing Complexity in Bali.* Princeton University Press, 2006.

Does age matter?

Subaks have existed at least since the eleventh century; today there are approximately a thousand.[4] By virtue of their location, upstream subaks can influence how much water reaches their downstream neighbors. Across the island, farmers recognize two management systems. In *tulak sumur* ("reject the wellspring"), everyone is free to plant whenever they like, which gives upstream farmers an advantage over their downstream neighbors, and usually leads to poor harvests. Alternatively, in *kerta masa* ("lawful/good timing"), irrigation schedules are chosen by consensus in subak meetings.

In an earlier study, we used neutral genetic markers to investigate the historical development of subaks on Bali.[5] Conventional archaeological methods can provide insights into the historical development of irrigation technology, but permit only indirect inferences about the social management of irrigation. But as we saw in chapter 2, variation in noncoding regions of the human genome that are subject to very rapid rates of mutation can be used to reconstruct the demographic histories of populations. In this way it is possible to trace micromigrations and changing patterns of relatedness within small communities in the recent past.

We considered two alternative scenarios for the expansion of irrigation and rice cultivation that would produce contrasting signals in the genetic structure of farming villages. If the expansion of irrigation was accomplished by the farmers rather than by their rulers, then population movements of men would occur only as a result of demographic pressure leading to the formation of new daughter settlements close to the parent villages. This budding model would predict the formation of small communities located along irrigation systems, with the oldest settlements located at the irrigation out-takes closest to the most ancient weirs or springs. Small population size and reproductive isolation would produce high rates of genetic drift; the older the community, the more evidence of drift. On the other hand, younger subaks would undergo a substantial founder effect, in particular for the male part of the population. Patrilineages should exhibit less evidence of movement on the landscape than matrilineages because only men inherit rights to

---

[4] The first appearance of the term subak is as the root of the word kasuwakan in the Pandak Bandung inscription of AD 1071. See R. Goris. *Inscripties voor Anak Wungçu. Prasasti Bali I–II. C.V. Masa Baru*, 1954.

[5] J. S. Lansing, T. Karafet, J. Schoenfelder, and M. A. Hammer. 2008. DNA signature for the expansion of irrigation in Bali? A. Sanchez-Mazas, R. Blench, M. Ross, et al., *Past Human Migrations in East Asia: Matching Archaeology, Linguistics and Genetics*. Routledge, pp. 377–95.

farmland. But matrilineages should also be very localized because of a strong cultural preference for marriage within one's own subak. Thus, a budding model of gradual irrigation expansion initiated by the subaks themselves would predict a clear pattern of population structure for both patrilineal and matrilineal relatedness in the rice-growing regions of Bali.

However, several authors have proposed an alternative scenario in which rulers managed the expansion of irrigation.[6] If so, none of these constraints would be in evidence. Instead, the population would serve as a reservoir of labor, which could be relocated by the princes to build and service new irrigation systems. Overall, this scenario would predict a more fluid population structure with less nucleation of settlements and less genetic drift within settlements compared with the budding model. In 2001, we gathered the genetic data needed to test these hypotheses.

We analyzed a total of 507 Balinese farmers from 21 subaks located in different regions of Bali.[7] Genetic samples were obtained from farmers in 8 subaks along the Sungi river ($N = 120$), as well as 13 subaks located elsewhere in Bali ($N = 287$), and a geographically distributed sample of 180 other Balinese men to provide context for the genetic patterns observed in the subaks. Analysis of these samples showed that older rice-growing villages typically have less genetic diversity than elsewhere on the island, reflecting the cultural preference for endogamous marriages within these villages, along with very low rates of migration into the community. Younger villages, or those with more immigration, show greater genetic diversity (Figure 7.1).

Overall, we found that subaks located furthest upstream on their respective irrigation systems demonstrate greater levels of genetic differentiation and diversity, suggesting that they were older than their downstream neighbors. The evidence from the Y chromosome was consistent with key features of this scenario: patrilocal residence, with very little movement on the landscape except for occasional micromovements to nearby daughter settlements. The older the subak, the more evidence for this pattern. Evidence from mtDNA was consistent with the contemporary observed pattern of patrilocal residence and preferential village or subak endogamy, but with occasional marriages outside the

---

[6] For a summary of this controversy, see J. S. Lansing and T.A. de Vet. 2012. The functional significance of Balinese water temples: A reply to critics. *Human Ecology* 40:453–67.

[7] Lansing, Karafet, Schoenfelder, and Hammer, DNA signature; T. M. Karafet, J. S. Lansing, A. J. Redd, et al. 2005. Balinese Y-chromosome perspective on the peopling of Indonesia: Genetic contributions from pre-Neolithic hunter-gatherers, Austronesian farmers, and Indian traders. *Human Biology* 77:93–114.

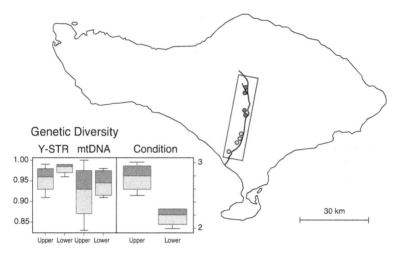

Figure 7.1. Map showing the eight subaks in the pilot study on the Sungi River of Bali. All four of the upstream subaks and two of the downstream subaks are small (mean size 67 ± 16 members), while the other two downstream subaks are much larger (231 and 535 members). Left histogram: mean Y chromosome STR genetic diversity: four lower subaks = 0.980, four upper subaks = 0.955; mean mtDNA diversity: lower subaks = 0.945, upper subaks = 0.923, $P = 0.031$. Right histogram: responses of 10 farmers in each subak to "What is the overall condition (state) of your subak?" Mean lower subaks = 2.15, upper subaks = 2.78, $P < 5.0 \times 10^{-10}$. Credit: Authors.

subak. Again, this pattern was most strongly evidenced in the oldest villages, and scales with time. There was a strong contrast between the close relatedness of families within subaks, compared to the background levels of relatedness in the whole population. These are not the patterns expected under the alternative scenario of state-controlled expansion of irrigation—that is, rulers transporting whole villages to newly constructed irrigation areas or, alternatively, bringing settlers from nearby villages.[8]

Intuitively it makes sense that the age of subaks would be related to their physical location, with the oldest subaks located furthest upstream. In the rainy season, rivers flowing down steep Balinese volcanoes are hard to control with traditional engineering techniques. Traditional weirs were made of earth, stones, and logs, and can be washed away by heavy rains. Consequently the oldest subaks would tend to be located upstream, where the smaller flows are easier to manage. One of the earliest dated Balinese

[8] J. S. Lansing, M. P. Cox S. S. Downey, et al. 2009. Robust budding model of Balinese water temple networks. *World Archaeology* 41:112–33.

royal inscriptions is kept in a temple near the headwaters of the Sungi River, and contains references to wet-rice agriculture and nearby streams (Babahan 1, AD 947).[9] Newer, larger subaks and irrigation systems are located along the lower stretch of the Sungi River; these subaks cannot be dated as precisely as the older ones, but were probably constructed beginning in the eighteenth century.[10]

A pilot study of cooperation

So, do the different demographic histories of upstream versus downstream subaks have consequences for their responses to social and environmental challenges? With the genetic data in hand, the obvious place to test this idea was in the eight subaks located along the Sungi River in the district of Tabanan, the largest rice-growing region in Bali (Figure 7.1).

We started with a simple pilot study of about ten farmers in each subak, who were also asked to play an experimental game that is often used to study cooperation. For the survey, about 10 randomly selected farmers in each subak were invited to answer a questionnaire. Farmers who participated were paid the equivalent of a laborer's day wage (30,000 Indonesian rupiah, or ~US$2.25). Survey questions were divided into two sections (Table 7.1). The first section asked farmers about their harvest yields for the past two years and whether they experienced losses due to water shortages or pest infestations. They were also asked how much farmland they owned or sharecropped over the same time period. The second section asked for the farmer's opinions about factors that might affect the subak's ability to respond to both social and environmental problems. These included the effectiveness of sanctions against norm violators; the ability of the subak to mobilize collective labor for maintenance of the irrigation works and the performance of temple rituals; the general condition or state of their subak, and its capacity to cope with either technical or social problems. The farmers were also asked about the effects of differences in caste and class on the functioning of their subak. Because rivalry between castes is endemic in Bali, the proportion of high versus low-caste members in each subak could affect its ability to respond to problems. Similarly, farmers were also asked to assess the effects of class differences (indexed by the proportion of sharecroppers versus landowners in the subak) on the efficacy of their subak.

[9] Goris, *Inscripties voor Anak Wungçu.*
[10] H. Schulte Nordholt. *The Spell of Power. A History of Balinese Politics 1650–1940.* KITLV, 1996.

Table 7.1

Survey questions. The 19 questions used in the reduced list for analysis are highlighted. See Appendix for details.

Question	Descriptor	Question	Descriptor
1	Own farmland	19	Pest damage in subak
2	Sharecrop land	20	Pest damage myself
3	Inherited a farm	21	Thefts of water
4	Purchase	22	Conflicts among members
5	Sold a farm	23	Choice of subak head
6	Income	24	Fines
7	Harvest	25	Crop schedule followed
8	Satisfaction with harvest	26	Plan work
9	Origin	27	Written rules followed
10	Condition of canals	28	Fines frequency
11	Condition of fields	29	Condition of subak
12	Synchronize	30	Decisions of subak accepted
13	Attendance at meetings	31	Technical problems
14	Participation in maintenance	32	Social problems
15	Attendance at ritual	33	Caste problems
16	Accept subak decisions	34	Class problems
17	Water shortages in subak	35	Resilience
18	Water shortages myself		

To facilitate comparisons of prosocial behavior both between subaks and with other published studies of the management of "common pool resources," the same farmers who answered our survey questions also played the Dictator Game. In this simple game, each player was given another day's wage and told that he could share as much, or as little, as he chose with another member of his subak who was present at the time. Players were assured that the identity of both givers and receivers would be kept confidential. Because of this guarantee of anonymity, data on offers in the game and genetic data were aggregated at the scale of subaks, while survey data were tabulated at the level of individual farmers. Survey results were analyzed using multivariate ANOVA and principal component analysis, which produced two clusters of subaks. Eigen-decomposition of the correlation matrices of these clusters clarified the harmonic well structure of each regime within the 11-dimensional principal component space.

Results of the pilot study

As we learned from our earlier study of the population genetics of the subaks, the genetic diversity of maternal and paternal lineages is significantly reduced in the four upper subaks relative to the four lower subaks ($P = 0.031$) (Figure 7.1). This suggested that the upstream subaks are more demographically stable; that is, they have a long history of continuous occupation by the same families. Farmers experienced both water shortages and pest damage to varying degrees. And as expected, water shortages were perceived to be a greater problem by downstream subaks, while pest infestations were perceived to be more problematic by upstream subaks ($P = 0.032$). This difference was also noted by the agricultural extension agents who assisted with the surveys.[11] In the most recent harvest recorded (2010), average rice harvests were slightly larger in the upstream subaks (5.4 tons/hectare) than downstream subaks (4.8), though this difference was not statistically significant. Subaks also experienced problems stemming from social conflicts, which may be affected by tensions arising from differences in either caste or social class. In the pilot survey, subaks varied in the proportion of their members who belong to the upper castes (0–24%). Comparative analysis of the effects of class differences were made possible because the mean proportion of owners versus sharecroppers (question 5, "ownership") is virtually identical for the two groups (upper subaks: mean 75.2%, SD 40%; lower subaks: mean 78.2%, SD 34%; $N = 83$).

Farmers in the upper subaks responded much more positively than downstream farmers to all survey questions, including the efficacy of sanctions; the ability of the subak to mobilize manpower and resources to carry out irrigation maintenance, perform rituals, and conduct meetings; and the overall condition and resilience of the subak (Pillai's trace of rank score = 0.49, $P = 2.9 \times 10^{-5}$). Principal components analysis of the eleven surveyed variables in the upper and lower subaks showed contrasting patterns (Figure 7.2). Overall, the responses of these eight subaks to social and environmental challenges fell into two contrasting patterns. The upstream and downstream subaks experience similar social and environmental conditions, but they respond to them in different ways. These

---

[11] J. S. Lansing, S. A. Cheong, L. Y. Chew, et al. 2014. Regime shifts in Balinese subaks. *Current Anthropology* 55:232–9.

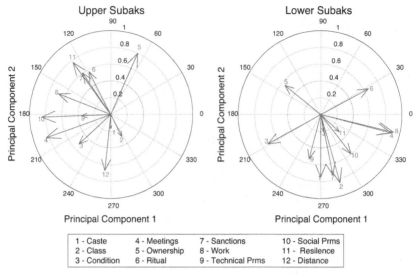

Figure 7.2. Biplots of upper and lower subaks. Left: Biplot for the upper four sub-aks, showing the correlation structure of responses to the surveys (PC1—85.5%, PC2—5.4%). Descriptor axes that extend beyond the equilibrium (inner solid) cir-cle are statistically significant. Most variables have different effects in the upper and lower subaks. For instance, in the upper subaks, there is no relationship between question 5, the proportion of sharecroppers vs. owners, and question 10, the ability of the subak to resolve social problems. Right: Biplot for the lower four subaks (PC1—92.0%, PC2—2.4%). Here "ownership" (question 5, the pro-portion of sharecroppers vs. farm owners) anticorrelates with question 10 (the ability of the subak to resolve social problems). Note that the axis of question 1 (effects of caste differences) has been pointed to 270° in both plots. The first two principal components account for nearly all of the variance in the correlation matrices. Credit: Authors.

differences only become apparent through a cross-scale comparison; they would be invisible to a study of either individual subaks or the system as a whole.

To explain these differences, size did not appear to matter: the two small downstream subaks resemble the large downstream subaks more closely than the small upstream subaks. Differences in genetic relatedness between the two groups are too small to affect cooperation directly (say, through a mechanism like kin selection), nor do they suggest the exclu-sion of outsiders. Instead, the differences are consistent with the view expressed by several upstream farmers that their subaks benefit from generations of shared history.

In ecology, the discovery of alternate stable states led to a shift from the investigation of equilibrium or near-equilibrium states to the study of stability boundaries for different regimes.[12] In this pilot study, we extended this approach to a fully coupled social-ecological system, revealing a clear separation between two regimes. The more successful upstream subaks flourish in a small, but deep, basin of attraction. Confident of their collective ability to meet any challenge, they are exceptionally cooperative. The less confident group of lower subaks also cluster around their own attractor, revealing that "muddling through" can also be a steady state, but with different dynamical relationships among state variables than in the upstream group.

Exploring differences in regimes

The simple questionnaire used in the pilot study was enough to suggest the existence of two regimes. A key discovery from the pilot study is that the significance of the farmer's answers to each question depends on the state of their subak: which regime is it in? Several questions naturally follow. Do other regimes exist elsewhere on the island? How stable are they? Can the survey data provide insights into their resilience? Today the subaks and their rice terraces are presently vanishing at a rate of about a thousand hectares each year. Could we use the survey data to explore the most probable transition paths between regimes, whether or not a transition is already under way?

To find out, we followed up the pilot study with a more comprehensive questionnaire of approximately 25 farmers in each of 20 geographically dispersed subaks. At the same time, we began to reconsider our analytical methods. If different regimes exist, is principal component analysis (PCA) the best way to search for them and analyze their dynamical interactions? This question is not confined to Bali, or indeed to social science: it arises in cases in which more than one unknown attractor may exist in a dynamical system. In the case of the pilot study, we solved it by performing PCA separately on the clusters of upstream and downstream subaks. At that point, the nearly linear correlations within each PCA became obvious. But suppose the underlying statistical manifold was more curvaceous? PCA belongs to a family of standard statistical methods that rely on a correlation matrix, either between all the questions in a survey, or all of the respondents. If the statistical distribution is the result of interactions in a linear dynamical system, PCA will capture them. PCA can be used to

---

[12] C. Folke. 2006. Resilience: The emergence of a perspective for social-ecological systems analysis. *Global Environmental Change* 16:253–67.

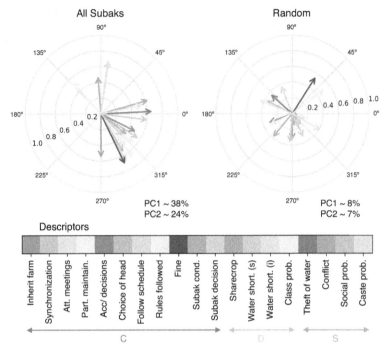

Figure 7.3. Comparison of PCA biplots of survey data from all 20 subaks and from randomized samples. Each descriptor is assigned a unique shade. The length of the arrow for each descriptor indicates its magnitude (contribution to the PCA). Arrows that are closer together are more correlated. Cooperative descriptors (C); defective descriptors (D); and social disharmony descriptors (S). Credit: Ning Ning Chung.

estimate the embedding dimension of attractors in nonlinear dynamical systems even when several attractors exist.[13] But if more than one attractor exists, as we have just seen, the resulting nonlinear relationships may be obscured.

With the new survey data from the additional 20 subaks in hand, we removed the relatively unimportant descriptors by means of higher-order clustering and analyzed the remaining 19 using principal component analysis. We observe three groups of closely correlated descriptors, which we term groups C, D, and S (see Figure 7.3). Group C contains correlated descriptors which depend directly or indirectly on the cooperativity of

---

[13] H. Kantz and T. Schreiber. *Nonlinear Time Series Analysis*. 2nd edition. Cambridge University Press, 2004.

the farmers. Group D is anticorrelated with group C and corresponds to defection. The relevant descriptors are mainly associated with problems such as limited water availability at both the individual and subak levels. The survey questions in group D also include the proportion of owners versus sharecroppers, and whether class differences (likely to be correlated with land ownership) affect noncooperative behavior. Group S is related to social disharmony and is observed to be uncorrelated with groups C and D. It has descriptors in the social domain such as social conflicts, social problems, caste, and frequency of water theft.

From clusters and attractors to regimes

How stable are these clusters? How strong is the evidence from the survey? And what might cause a subak to transit from one cluster to another? To find out, we add a new layer of analysis to the PCA by adapting the concept of an "energy landscape" from statistical physics. This enables us to quantify the strength of the attractors, and their proximity to each other, based on the PCA.

To construct the energy landscape, we have to fit the relative frequency of a set of subak states to a Boltzmann distribution. These subak states are coarse-grained states of each subak defined by its principal components. The energy landscape is formed over these subak states, with its topology directly proportional to the state's density (Figure 7.4).[14] The strength of the attractor for a given configuration of Principal Components in a cluster of subaks is defined by its density, which appears as its depth in the energy landscape: the denser the state, the greater the depth. As the density of a state weakens, the depth decreases. In this way, the energy landscape provides a visual representation of the strength of the attractors and their basins of attraction, which define the three regimes discovered by the subak-level PCA. Three clearly distinct basins emerge in the energy landscape (Figure 7.5, bottom). The dominant Attractor C, with 16 subaks, has an energy of −3.07 (in arbitrary units), compared to higher energy, less stable Attractors A (−0.52, 2 subaks) and B (−0.52, 2 subaks). These basins reflect regions of increased stability, which we interpret as regimes, and their relative depths indicate their relative stability.

With the energy landscape determined, transition paths between attractors can be calculated. The idea here is similar to the evaluation of transition paths between metastable states (attractors) in chemical kinetics and protein folding problems. Typically, such transitions are driven by noise, which enables the system to overcome an energy barrier as it

[14] The density of a state is defined as the number of subaks per state.

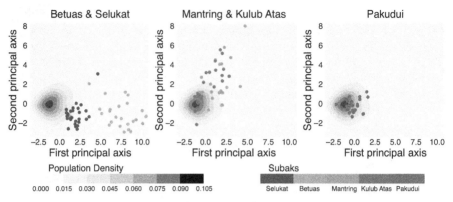

Figure 7.4. Combined density plot for 20 subaks. Aggregate responses from 15 subaks are tightly clustered in a single attractor basin, located near the bottom left in each figure (Attractor C). Shaded dots show responses from individual farmers from subaks that fall outside this attractor. Two weak attractors are evident: Betuas and Selukat (Attractor A), and Mantring and Kulub Atas (Attractor B). Pakudui falls within Attractor C despite a recent social shock, which we interpret as a sign of high resilience (see text). Credit: Ning Ning Chung.

transits between the two attracting states. In our case, the noise arises from the variability of the social and ecological conditions. Instead of reaction coordinates as in the protein folding problem, our collective variables are the dominant principal components of the system. The transition path is the minimum energy pathway, which is also the most likely path between the attractors. We hypothesize that the transition pathway is the one that tracks the smallest difference between the variables that dominate the three principal components. The survey data provide us with the boundary conditions for such hypothetical transitions between regimes to unfold. We estimate that the descriptors that dominate along the transition paths are indicated by the absolute difference between the mean descriptor state of the two attractors (see band in Figure 7.5). On this basis, two contrasting paths between the cooperative Attractor C and the other two attractors emerge. Social problems dominate the pathway from B to C, and are negligible for A to C. These balance out for A to B. Water availability dominates the transition path from C to A, because it is a constant problem for C subaks but essentially absent for A.

These predictions can be evaluated by taking a closer look at the subaks in the three attractors. As Tolstoy observed in *Anna Karenina*, while happy families are all alike (as in Attractor C), every unhappy family is unhappy in its own way. Subaks Betuas and Selukat (Attractor A) are located near the sea, near the terminus of their irrigation systems, but

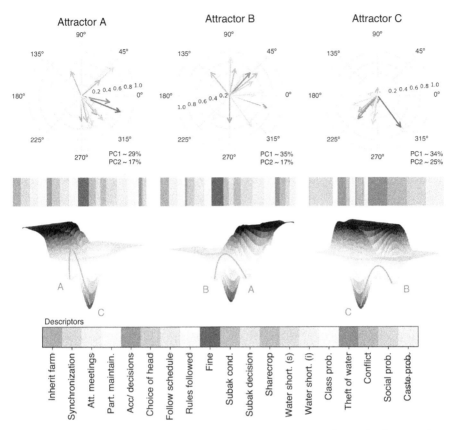

Figure 7.5. Three attractors and the transition paths between them. The first panel shows the biplots for the three attractors. Attractor A includes subaks Betuas and Selukat; Attractor B consists of Mantring and Kulub Atas; all other subaks are in Attractor C. The second panel shows the change in each descriptor along the transition path. The third panel shows the energy landscape and the transition path. Credit: Ning Ning Chung.

nonetheless have abundant water thanks to eleven natural springs. Indeed these two subaks scored highest on satisfaction with water availability. But in 2002 it became known that a coastal highway would go through their land, and speculators began to buy up subak land in anticipation of the construction of the highway. After the road was completed, many farmers leased their own land back from the speculators, and so became sharecroppers. The highway bisects both subaks, and the heads of both subaks said that the subaks are now in danger of collapse. In the survey, farmers described the condition of their subaks as poor in Betuas (mean response 2.85) and fair in Selukat (3.29). The mean response for "condition of my subak" for all subaks was 3.82.

Subaks Mantring and Kulub Atas (Attractor B) cope with different problems. Both are located further upstream along their respective rivers. We confirmed the survey results indicating that the main irrigation canal of Kulub Atas needs repair, water shortages are frequent, and the head of the subak is unpopular. The overall condition of Kulub Atas was rated 3.64 by the farmers. Mantring had the lowest harvest of the 20 subaks, and their farmers scored a below-average overall condition of the subak at 3.70. This assessment was further exacerbated by low scores in satisfaction with harvests, high social problems, frequent water theft, irrigation canals in poor repair, and poorly synchronized irrigation schedules.

Cryptic regimes?

The three regimes are each characterized by a single attractor. More information can be obtained by exploring variation in the responses to the surveys within each attractor. To do so we utilize Fisher Information, a measure of the extent to which the survey responses of an individual farmer, projected onto PC-space, is predictive of the survey responses of other farmers in the same subak. Most subaks fall into Attractor C, with high Fisher Information, which suggests that the survey responses reflect consistent patterns of dynamical relationships. This phenomenon—similarity in survey responses between individuals mirroring cooperation—is striking, and consistent with a canonical prediction from theoretical game theory,[15] which argues that population structure—a tendency for individuals to interact with other individuals who are more similar in their behavior—can promote the evolution and persistence of cooperation in a population.

But what of the other attractors? Attractors A and B have low Fisher Information, which means that their internal distribution of responses to survey questions are both widely scattered and different from those of the subaks in the main cluster (C). Figure 7.4 shows the distribution of individual farmers in villages in the noncooperative regimes A and B, and a single village which lies in the cooperative regime C. Farmers in regimes A and B show considerable heterogeneity. We observe a clear trend whereby

[15] T. Killingback, J. Bieri, and T. Flatt. 2006. Evolution in group-structured populations can resolve the tragedy of the commons. *Proceedings of the Royal Society B* 273:1477–81; F. C. Santos and J. M. Pacheco. 2006. A new route to the evolution of cooperation. *Journal of Evolutionary Biology* 19:726–33.

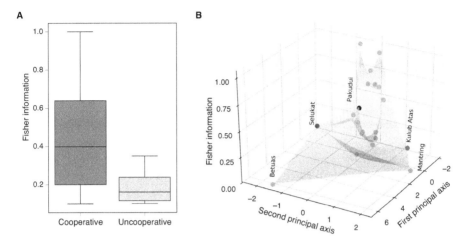

Figure 7.6. Fisher Information of survey responses in the 20 subaks. For each subak, we fit the distributions of its first and second principal components to either the Gaussian, Rayleigh, or Pareto distributions. Fisher Information of each of the two principal components is then calculated according to the fitted distribution. The sum of the Fisher Information contained in the first and second principal components is recorded. A. Distribution of Fisher Information in cooperative and noncooperative subaks, B. Plotting Fisher Information on the 2-dimensional PCA space reveals a Fisher Information landscape documenting cohesion within the different subaks. Credit: Ning Ning Chung.

subaks in the noncooperative regimes tend to have lower Fisher Information than those in the cooperative regime (Figure 7.6A). Despite the small numbers of farmers, this tendency is statistically significant (Welch's t-test one-tailed $P = 0.006$).

We also observe that while most of the cooperative subaks show extreme homogeneity (e.g., Kedisan Kaja and Kebon), others, still deep within Attractor C, include surprising variability (Figure 7.6B) (e.g., Dukuh and Tegan). Probing the survey results for these two subaks, we observe that Dukuh is plagued by a greater level of uncooperative issues while Tegan is more adversely affected by poor environmental conditions in comparison to the other cooperative subaks. Both of them also face the social problem of more frequent water thefts. Tellingly, farmers in these two subaks describe the role of democracy in the governance of their subak as a "veneer" rather than an actuality, at higher rates than the other cooperative subaks. This raises a question: are these two subaks potentially less resilient than would be predicted from the energy landscape results alone? Is it possible to infer the latent potential for transitions on the adaptive landscape that have not yet begun?

A clue that this may be possible is provided by another subak that lies deep in the cooperative regime (C). Analysis of the Fisher Information as well as its location on the energy landscape suggests that this subak, Pakudui, is more resilient than the raw questionnaire data suggest. The overall pattern of reponses situates Pakudui in the cooperative regime (Figure 7.4). But the response to Question 33 (caste problems) did not fit this pattern. This question asked, "In your opinion, is there a connection between the capability of the subak and caste conflicts within the subak?" Farmers had three choices: Frequently (scored as 1), sometimes (scored as 2), and seldom (scored as 3). The mean response for all subaks was 2.94, but for Pakudui it was 1.88. This anomaly led us to revisit the subak to inquire about caste. We learned that there has been a long standing dispute between two groups in the village, numbering 60 and 15 households, about their caste prerogatives. The origins of the conflict go back to a dispute that began in the 1960s about the management of income from the sale of rice belonging to a village temple. One group claimed that they were exempt from the responsibility to contribute to the annual ritual cycle at the temple, because they should be credited with the income from the temple's ricelands. This dispute quietly simmered for decades, but heated up the year before our survey when one group refused to allow a member of the other group to be buried in the cemetery for two days until police intervened.

What's interesting about this result is that this severe social conflict apparently did not cause the subak to become dysfunctional, even though members of the two groups barely speak to one another. Yet responses to the other survey questions were clustered within the cooperative regime (Figure 7.4), and the Fisher Information fell within the middle range (Figure 7.6B). We interpret this result as reflecting the resilience of subak Pakudui to internal stress. Pakudui is located at a relatively high elevation, in the upper reaches of the Petanu River where some of the oldest evidence for irrigated rice is found. It is one of 15 subaks that form the congregation of a major regional water temple, Masceti Pamos Apuh. An inscription provisionally dated to the twelfth century mentions contributions made by the irrigation leaders of several of the subaks belonging to this water temple (Sebatu, Kedisan), though Pakudui is not mentioned by name. This evidence suggests that Pakudui is an ancient and demographically stable subak. Informants in the village commented that despite their differences, families in Pakudui appreciate the need to prevent quarrels from spilling over into the subak, lest everyone suffer.

Conclusion

Here for the first time, we were able to objectively detect and explicitly quantify the resilience of the subak system. We suggest that these results from a simple questionnaire may have relevance beyond the Balinese case. From a statistical perspective, the existence of multiple regimes implies clustering. While there are many statistical methods for detecting clusters, little attention has been given to the question of how clusters arise in social-ecological systems, how they change, and transitions between them. Once we detected the clusters present in the PCA of the subaks, we were able to analyze the dynamic relationships between variables in each cluster (regime), which we might think of as traits. Thus, we can think of responses to survey questions as combining to define behavioral tendencies in regimes, such as attitudes toward the environment or attendance at meetings. A change in a trait might be destabilizing in one regime but have little consequence in others. This realization led us to search for threshold-dependent traits that could trigger transitions between regimes, like the "veneer of democracy" or the salience of caste conflicts. While a standard statistical analysis might flag these as problems, the energy landscape and Fisher Infomation results provided a context from which it became possible to evaluate their relationship to the resilience as well as the internal dynamics of each subak. Simply put, the significance of the farmer's answers to each question depends on the state of their subak, and our methods make it possible to quantify that dependency.

By analyzing the subak system at hierarchical scales that follow its natural structure, in the second survey we discovered three regimes. The largest regime consists of subaks in which responses to all survey questions are relatively cohesive and cooperation is the norm. In addition to this expected cooperative regime, we discovered two less functional attractors facing different sets of challenges, with a tendency toward either endogenous problems (Attractor B) or exogenous disruption (Attractor A). Such variation is probably a natural property of coupled social-ecological systems, one that becomes clearly visible in the attractor basins of the energy landscape (Figure 7.5). In coupled socioecological systems like subaks, where both social and ecological components may interact nonlinearly, cryptic regimes may be widespread. Identifying such regimes, assessing their resilience, and predicting transition paths between them is an important step toward understanding and protecting the socioecological systems on which we depend.

CHAPTER 8

From the Other Shore

It is the solution of the riddle of history and knows itself to be the solution.

—*Karl Marx*

If there were a libretto, history would lose all interest.

—*Aleksandr Ivanovich Herzen*

"Time is a mental concept," said Pringle. "They looked for time everywhere else before they located it in the human mind. They thought it was a fourth dimension. You remember Einstein."

—*Clifford D. Simak*[1]

Metatheory (a theory whose subject matter is some other theory) is a word for philosophers, not fieldworkers. But without it our book lacks a conclusion, so we will gently extract it from its home in philosophy, and explain why we feel that we need it here. It all comes down to chance.

For the past two centuries, chance has been anathema to the social and historical sciences. When Darwin introduced chance in the early editions of his writings on evolution, it caused a firestorm of criticism, drawing hostile reactions even from Darwin's scientific supporters.[2] Today, the role of chance in phenomena such as convergent evolution has reignited as one of the most controversial questions in evolutionary theory, and we are cognizant of the need to approach this topic with caution. Still, we take courage from the example of Alfred Russel Wallace. In 1858, toward the end of his "eight years' wanderings among the largest and the most luxuriant islands which adorn our earth's surface," Wallace wrote a letter to Charles Darwin "in which I said that I hoped the idea [of selection on differences that arise by chance] would be as new to him as it was to me, and that it would supply the missing factor to explain the origin of species." Wallace sent his letter on the mail steamer from the eastern

---

[1] K. Marx, *Economic and Philosophical Manuscripts* of 1844; A. I. Herzen, *From the Other Shore*, 1850; C. D. Simak, *Time and Again*, 1951.

[2] C. Johnson. *Darwin's Dice: The Idea of Chance in the Thought of Charles Darwin.* Oxford University Press, 2014, p. 111.

Indonesian island of Ternate in April 1858, enclosing a brief handwritten essay in which he concluded:

> This progression, by minute steps, in various directions, but always checked and balanced by the necessary conditions, subject to which alone existence can be preserved, may, it is believed, be followed out so as to agree with all the phenomena presented by organized beings, their extinction and succession in past ages, and all the extraordinary modifications of form, instinct, and habits which they exhibit.

Darwin received the letter on 18 June, and wrote to the president of the Royal Society the same day asking what he should do with Wallace's work. With Charles Lyell's encouragement, Darwin quickly prepared a lecture to the Royal Society proposing a theory of evolution by natural selection.

We hasten to say that we are under no illusion that our "islands of order" are remotely comparable to that discovery. But islands of order (attractors)—and the transition paths between them—offer new tools to analyze Wallace's "progression by minute steps in various directions." So far, we have used these tools on a case-by-case basis. Evolution creates branching processes, producing (as Darwin wrote) "endless forms most beautiful." But from a modeling perspective, the "checks and balances of necessary conditions" constrain branching processes and steer them to attractors. In each of the chapters of this book, we have modeled such processes to discover their attractors. Presently each model stands alone, designed to answer a specific question. But the questions are often related. Consider kinship: in successive chapters we have traced its effects on colonization, genetics, social dominance, language, and cooperation. An obvious next step is to consider whether and how these models may be related. In this chapter, we will explore that possibility by shifting our attention from the numerical outputs of the models to their dynamics.

The metatheoretical context emerges for this reason: to constrain branching processes with models is a way to box in chance, and thereby gain insight into fragments of social life. But "chance" has many meanings,[3] and choosing a model obliges us to pick one. We can compare models that address related questions, but each encodes a specific

---

[3] See C. Johnson. *Darwin's Dice*, 24: "The following list of usages/meanings is not intended to be exhaustive or systematic but rather to illustrate how extensive the idea has become in the days since Darwin. In particular I make no mention here of the 'chances of survival' idea discussed in the second part or of the 'chance survival of neutral variations' idea that has emerged in the neutralist literature.... Beatty (1984, 186–7) discriminates among 'chance' as 'undesigned,' 'chance' as the Aristotelian idea of 'coincidence,' and 'chance' as the Laplacean idea of 'cause unknown,' and contrasts these with Darwinian 'chance (i.e., random) variation' and the more recent idea of

process. To link them is conceptually simple but technically challenging. Still, doing so has the potential to bridge the gap between our models and the theoretical and historical questions posed by anthropologists. On one level these concern the social realities created by interacting processes that we have addressed, like marriage rules, language transmission, and competition, for social rank. But at a deeper level of abstraction, they concern change itself.

In the 1930s, a discourse about change emerged among the anthropologists working in the eastern archipelago. The key question was whether migrations alone could explain the observed distribution of languages and social structure. Strongly patterned variation in the retention of many features of social structure, language, and culture across the archipelago challenged a Wallacean model based on simple drift and migration. New questions were proposed, and the ensuing debate persisted for decades among Dutch, French, British, Indonesian, Australian, and American anthropologists. Our first goal in this concluding chapter is to explore how far our models can take us toward answering those questions.

The chapter is organized as follows: we begin by revisiting the Austronesian expansion, and review the theoretical debates that ensued in anthropology. Next we revisit our models. In previous chapters, we used phase portraits simply to illustrate their dynamics. Here we make active use of phase portraits, individually and in interaction, to explore the relevance of our empirical and analytical models to anthropologists' conceptual models of change.

Revisiting the Austronesian expansion

We began this book with Joseph Banks's discovery that the languages of the Pacific are systematically related. Banks suggested that the cause was migration, an idea that was taken up a century later by Wallace, who concluded that "[i]t is undoubtedly true that there are proofs of extensive migrations among the Pacific islands, which have led to community

---

'chance' as 'random drift' (cf. Shanahan, 1991). Eble (1999, and discussion in Millstein, 2000) identifies five meanings of chance in evolutionary biology, discriminating, importantly, between 'chance as ignorance of causes' (a Laplacean idea) and 'chance as random with regard to future adaptive needs' (the Darwinian idea under scrutiny here). Millstein (2011, 245–6) gives a 'list' (with explanations) of seven 'meanings of chance' in evolutionary theory as of the early twenty-first century. She traces only three of these back to Darwin. But these scholars agree that Darwin is responsible for opening the possibility that 'chance' could be a factor in evolution at all."

of language from the Sandwich group [Hawai'i] to New Zealand."[4] Although Wallace's *Malay Archipelago* was mostly about the natural world, he devoted the conclusion to "the races of man in the Malay Archipelago," which he attributed to separate migrations. A half century later, J.P.B. de Josselin de Jong revisited the question of migration in his inaugural lecture as head of the anthropology program at Leiden University, the leading center for Indonesian research. De Josselin de Jong argued that because the peoples of the archipelago share a common ancestry, to understand their languages, cultures, and social organization, it would be necessary to situate them within a comparative historical context. This led him to propose "The Malay Archipelago as a Field of Ethnological Study," an idea that set the agenda for the Leiden school of anthropology.

While acknowledging that migrations had taken place, de Josselin de Jong suggested that historically related societies should be investigated by comparative studies of their *structural core*, which he defined as shared features of social structure that are sufficiently different to make fruitful comparison possible. Minor differences in the structural core of neighboring societies imply the existence of a *protosystem*, which existed in the past but is no longer intact.[5] Influenced by this phylogenetic perspective, Leiden anthropologists began to speculate that superficial variations in social structure—for instance, patrilineality versus matrilineality—might not be the result of successive migrations as previously thought.[6] Instead, it was argued that Austronesian societies share a core set of ideas and institutions that find expression in cognitive classificatory systems or *structures* linking social organization, cosmology, and myth. In 1935, F.A.E. van Wouden surveyed the ethnographies of eastern Indonesia and concluded that over the whole region, "in spite of the extreme unilineal character of the descent systems, both patrilineal and matrilineal descent are yet taken into account."[7] As Gottfried Locher observed in 1968, "the great advance in understanding effected in the thirties was primarily the idea that accentuated matrilineal grouping, similarly marked patrilineal

---

[4] A. R. Wallace. *The Malay Archipelago*. Macmillan, 1868, pp. 454–5.

[5] "The deeper we delve into this system, the clearer it shows itself to be the structural core of numerous ancient Indonesian cultures in many parts of the Archipelago. The organization of these archaic societies is closely connected with, and even to a large extent dominated by, the kinship and marriage system." J.P.B. de Josselin de Jong. *De Maleische archipel als ethnologisch studieveld*. J. Ginsberg, Leiden, 1935.

[6] F.A.E. van Wouden. *Types of Social Structure in Eastern Indonesia*. Martinus Nijhoff, 1968, p. 153 (original 1935).

[7] Ibid.

grouping, and double unilineal grouping could belong to one and the same structure."[8]

In early twentieth-century Indonesia, as van Wouden observed, the patrilineal principle dominated. But as he further noted, an underlying dualistic principle implies that female origins and descent remain significant. Social groups composed of related individuals need to form and retain alliances and to commemorate their origins. For those purposes "it is absolutely immaterial whether the principle of genealogical grouping is matrilineal or patrilineal."[9] Instead, "one of the most striking facts is that in almost every one of the larger regions into which the area may be divided there is a people who are sharply distinguished from their patrilineal neighbors by their matrilineal descent groups." Van Wouden rejected the hypothesis of multiple migrations, observing that "the various cultures in question exhibit too great a homogeneity to make it necessary to resort to migration-hypotheses."[10]

Subsequent ethnographic research in the archipelago supported van Wouden's conclusion. For example, at the far western edge of Island Southeast Asia, communities on the island of Nias are organized as *banua* (villages) consisting of exogamous patrilineages.[11] Four thousand kilometers to the east, at the eastern border of the archipelago on the islands of Tanimbar, villages consist of rows of named and unnamed houses (*uma*) linked by matrilateral alliances and affinal relations.[12] In both of these societies, and in many others, a concept of cosmological

[8] G. W. Locher. Introduction to the English translation in van Wouden, *Types of Social Structure.*

[9] Van Wouden, *Types of Social Structure*, 153.

[10] "In eastern Indonesia an understanding of the house embraces more than its physical structure and the symbolic significance attached to its parts. The house defines a fundamental social category. House structures are particular local representations of this wider conception. They define what is generally regarded as a 'descent group' but might more appropriately be referred to, in Austronesian terms, as an 'origin group.' This group is of a variable segmentary order. It provides a sliding scale that may be associated with different physical structures depending on the development of the group, its conception of its origin and its relation to other groups.... Generally the societies of eastern Indonesia possess a category that defines a social group larger than the house. 'Houses'—often with specific ancestral names—make up units within the clan." J. J. Fox. 1993. Comparative postscript. In J. J. Fox, ed. *Inside Austronesian Houses: Perspectives on Domestic Designs for Living.* Research School of Pacific and Asian Studies, Australian National University, p. 170.

[11] E.E.W.G. Schröder. *Nias: Ethnographische, Geographische en Historische Aantekeningen en Studien.* Vol. 2, E. J. Brill, 1917.

[12] S. McKinnon. 1995. Houses and hierarchy: The view from a South Moluccan Society. In J. Carsten and S. Hugh Jones, eds. *About the House: Lévi-Strauss and Beyond.* Cambridge University Press, p. 175.

dualism is expressed in the complementarity of father (*ama*) and mother (*ina*). Houses are ranked according to their distance from an origin, and the contrast between older and younger permeates the kinship systems, social precedence, and cosmological myths of origin.[13] These attributes of Austronesian societies sharply contrast with the social organization of neighboring Papuan societies, which as archaeologist Peter Bellwood observed, "seem to lack totally not only houses but also any concept of genealogically based ranking, whether of persons or descent groups."[14]

The protosystem and the structural core

By now, the work of many ethnologists has documented the existence of various features of de Josseyln de Jong's structural core across the Malay archipelago. Empirically, the structural core consists of a collection of shared concepts, often expressed as contrastive principles such as mother/father or older/younger. Bellwood suggests that they can be understood as reflecting a "founder-focused ideology," which includes reverence for founding ancestors (both genitor and genitrix), the naming of groups of descendants after them, and ascription of social rank in relationship to their origins, which influences or determines rights to land, labor, and ritual prerogatives.[15] Avoiding the inevitable decline in status as lineages proliferate and grow distant from their founders would provide a plausible motive for junior members of society to seek opportunities to become founders of new communities. Some such explanation appears to be needed given Bellwood's rough estimate of the speed of the

[13] P. Bellwood, J. J. Fox, and D. Tryon. *The Austronesians: Historical and Comparative Perspectives*. Research School of Pacific Studies, Australian National University, 1995, pp. 214–5.

[14] P. Bellwood. 1996. Hierarchy, founder ideology and Austronesian expansion. In J. J. Fox and C. Sather, eds. *Origins, Ancestry and Alliance: Explorations in Austronesian Ethnography*. Research School of Pacific and Asian Studies, Australian National University, p. 22.

[15] "Considered comparatively, ideas of origin may vary significantly among Austronesian populations but these ideas generally rely upon a combination of elements often phrased in terms of common metaphors based on recognizable cognate expressions. It is this discourse on origins that is distinctively Austronesian. It is possible, in linguistic terms, to trace the use of cognate terms among different groups but much more is involved in this discourse. Frequently similar metaphors of origin persist even when the terms used in these metaphors are unrelated." J. J. Fox 1996. Introduction. In J. J. Fox and C. Sather, eds. *Origins, Ancestry and Alliance: Explorations in Austronesian Ethnography*. Research School of Pacific and Asian Studies, Australian National University, p. 4.

initial Austronesian expansion—about 75 km per generation, speeding up to ~325 km into Melanesia and western Polynesia.[16]

Still, the great age of Austronesian migrations ended over a thousand years ago. How then to explain the persistence of a founder-focused ideology as the structural core of Austronesian societies today? In Leiden, de Josselyn de Jong's original notion that a protosystem once existed at the dawn of the Austronesian expansion gradually gave way to the concept of the structural core as an ideology encoded in a cognitive system, for which the passage of time is essentially irrelevant. The structural core became defined as the idea that "cosmos and society are organized in the same way," by means of a core set of binary symbolic oppositions (male/female, older/younger, trunk/tip, cosmos/society). As Rodney Needham, an Oxford professor of anthropology, wrote in 1968, this core set of structural principles forms a "scheme of social categories... [that] serves as the model for an all-embracing classification."[17] A question arose as to whether the retention of these structural principles required the retention of the Austronesian words that define them. On this point, thirty years later anthropologist James Fox supplied this explanation: "Considered comparatively, ideas of origin may vary significantly among Austronesian populations but these ideas generally rely upon a combination of elements often phrased in terms of common metaphors based on recognizable cognate expressions. It is this discourse on origins that is distinctively Austronesian. It is possible, in linguistic terms, to trace the use of cognate terms among different groups but much more is involved in this discourse. Frequently similar metaphors of origin persist even when the terms used in these metaphors are unrelated."[18]

Thus, the structural core of Leiden anthropology came to be interpreted as a cognitive system that is defined by its power to annul change (in the Wallacean sense of change as random drift). The implication is that over the centuries, while the words used to define these symbolic contrasts inevitably drift apart, the all-embracing system of cognitive classification that they encode remains intact. Consequently, in the words of Claude Lévi-Strauss, "it is not the Leiden anthropologists but the Indonesians who are the great structuralists!"[19] Lévi-Strauss took the argument against drift a step further by proposing a theory

[16] Bellwood, Hierarchy, founder ideology and Austronesian expansion, 19.

[17] R. Needham. 1968. Introduction to the English translation of F.A.E. van Wouden, *Types of Social Structure in Eastern Indonesia*. Martius Nijhoff, p. 2.

[18] Fox, Introduction, 4.

[19] R. De Ritter and J.A.J. Karremans. 1988. *The Leiden Tradition in Structural Anthropology: Essays in Honor of P.E. de Josselin de Jong*. Brill, p. 31.

that this structural core is an example of a hitherto unrecognized form of social structure intermediate between the elementary and complex structures that he had previously distinguished in 1949. Inspired by van Wouden, Lévi-Strauss described this structure as a "house society" (*société à maison*). In a series of lectures at the Collége de France in 1978, Lévi-Strauss defined house societies by contrasting them with lineage-based social systems. His initial inspiration for these *sociétés à maison* was the noble houses of Europe: the historical house of Plantagenet or the fictional house of Usher. Lévi-Strauss observed that houses may appear in hierarchical societies as durable social groupings, which "reunite or transcend" opposing categories such as descent/alliance, patrilineality/matrilineality, hypergamy/hypogamy, and close/distant marriage.[20] Thus, Lévi-Strauss generalized van Wouden's analysis of Eastern Indonesian social structure as a phase in the evolution of complex societies.

Lévi-Strauss's proposal that the Austronesians had introduced a particular form of house society to Island Southeast Asia was first addressed by the Leiden School anthropologists in 1987 in a collection of essays. Subsequently the idea was debated in numerous articles and edited volumes.[21] In 1993, James Fox and collaborators offered a historical perspective on Austronesian *sociétés à maison* based on a comparison of contemporary ethnographic studies across Island Southeast Asia, coupled with historical linguistics. As Fox noted, the reconstructed lexicon of Proto-Austronesian contains the word **Rumaq*,[22] which Robert Blust glosses as a descent group or house.[23] A second relevant term is Proto-Malayo-Polynesian **banua/*panua*, a more polysemous word whose glosses include inhabited territory, homeland, community, and land-owning kin group.[24] Blust also identified a large number of Proto-Austronesian terms for

[20] C. Lévi-Strauss. *Le Regard Éloigné*. Plon, 1983. (Published in English as *The View from Afar*.)

[21] C. MacDonald. 1987. Sociétés "à maisons" et types d'organization sociale aux Philippines. In C. MacDonald, ed. *De la hutte au palais: sociétés "à maisons" en Asie du Sud-Est Insulaire*, CNRS Press, pp. 67–87.

[22] In historical linguistics, an asterisk indicates a reconstructed, but unattested, ancestral word.

[23] R. Blust. 1980. Early Austronesian social organization: The evidence of language. *Current Anthropology* 21:205–47.

[24] R. Blust. 1987. Lexical reconstruction and semantic reconstruction: The case of Austronesian house words. *Diachronica* 4:79–106; A. Pawley. 2005. The meaning(s) of Proto Oceanic **panua*. In C. Gross, H. D. Lyons, and D. A. Counts, eds. *A Polymath Anthropologist: Essays in Honour of Ann Chowning*. Research in Anthropology and Linguistics, Monograph No. 6, University of Auckland, Department of Anthropology, pp. 211–23.

the physical architecture of the Malayo-Polynesian house.[25] Commonly, Austronesian houses define social groups and connect them to the past, using a symbolic vocabulary that emphasizes origins and founder rank. Typically, as Fox notes, the house is regarded as the ancestral embodiment of the group it represents, engaged in marital and affinal alliances with other houses.[26]

We suggest that these concepts of change—protosystem, structural core, and house—offer complementary insights, but are not mutually consistent. De Josselyn de Jong's protosystem models change as drift, but he offers no explanation for its origin or persistence. The structural core retains the concept of the protosystem, but removes it from drift. Its selective retention or persistence is explained by its grip on the social imagination, but this concept also fails to account for its origins. Lévi-Strauss's *sociétés à maison* situate a particular structural core (the house model) in a progressive or developmental sequence of types of social organization, but the dynamics producing that progressive change are not addressed. Considered separately, our models do not appear to have direct implications for resolving the tension between these concepts of change. In the next section, we reanalyze them with this challenge in mind.

Phase portraits

We are not the first to ask how to connect slowly evolving historical currents with events in the recent past. This question was famously addressed by the historian Fernand Braudel in 1958, and revisited in 1987.[27] Braudel's influential solution was to give priority to the patterns that originate at the dawn of major historical epochs (for Braudel, the ancient Mediterranean). Like us, Braudel interprets the present in light of the distant past. He argues that events of the recent past should be interpreted in light of enduring patterns that shape them, the *longue durée*. This perspective led him to view events as the ephemera of history, which

---

[25] Blust's reconstruction points to a raised structure built on "posts," entered by means of a "notched log ladder," with a "hearth," a "storage rack above the hearth," "rafters," and a "ridge-pole covered in thatch." R. Blust. 1976. Austronesian culture history: Some linguistic inferences and their relations to the archaeological record. *World Archaeology* 8:19–43.

[26] J. J. Fox. 1993. Comparative perspectives on Austronesian houses. In J. J. Fox, ed. *Inside Austronesian Houses: Perspectives on Domestic Designs for Living.* Research School of Pacific and Asian Studies, Australian National University, pp. 1–28.

[27] F. Braudel. 1958. Histoire et sciences sociales: La longue durée. *Réseaux* 27:7–37.

"pass across its stage like fireflies."[28] He describes the *longue durée* as the longest conceivable historical temporality, observing that "everything gravitates around it."

We invoke Braudel here because his use of models forms a counterpoint to ours and articulates a different concept of historical time. For Braudel, the *longue durée* provides the context needed to interpret events. The purpose of models is to discover large causes and central tendencies. As the data pile up, it becomes possible to turn the hour-glass upside down, "from the event to the structure and then from the structures and models back to the event."[29]

But turning the hour-glass upside down is not an option for us. Our first model, in Chapter 2, offered an explanation for the genetic barrier along the Wallace line and the spread of Austronesian languages, based on a slight tendency for Austronesian women to accept husbands from neighboring Papuan communities. The model can be described in a phase portrait (Figure 8.1), which shows progressive changes in the male and female ancestry of islanders moving eastward from the Wallace line. We initially created it to explain the distribution of languages and genetics in a handful of villages in Central Timor, and later extended it to encompass the initial phase of the Austronesian colonization of the Pacific. It is thus simultaneously a model of the *longue durée* of the Austronesian expansion, extending over tens of generations and half the globe, and a model of recent events in Timor. It differs from the Braudelian concept of a historical model in two ways. First, it does not seek to discover central tendencies in aggregations of data from many sources. Instead, it captures change. Turning the hour-glass upside down would reveal not powerful gravitational forces, but rather the faint traces of interethnic love stories, beginning at the dawn of the Austronesian expansion. Second, we followed up this model not by adding more details and parameters, but by pursuing its implications, in particular by looking more closely at the genetic signatures created by marriage practices. That was the main theme of Chapter 3, where we learned that across many islands, the genetic diversity of male and female ancestors at the scale of villages is far from uniform. At first, we explored the possibility that these differences could be the result of competition for mates, driven by selection for dominance. But this explanation proved to apply only to a small fraction of villages.

---

[28] F. Braudel. 1995. *The Mediterranean and the Mediterranean World in the Age of Philip II*. University of California Press, Vol. II, p. 901.

[29] Braudel, Histoire et sciences sociales, 751.

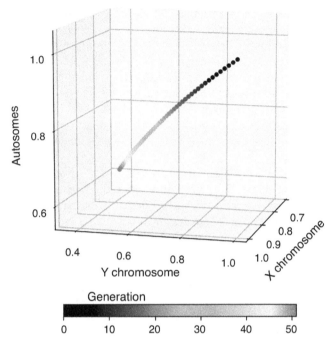

Figure 8.1. Phase portrait of the house model as a protosystem. The model predicts the genetic consequences of a marriage system in which an average of 2% of women in a matrilocal Austronesian village marry a Papuan man. Each shaded point describes the genetic consequences in a single generation. As time proceeds, the points progress from upper right in the first generation to lower left in the 50th. The model further predicts that the children of these matrilocal marriages will speak an Austronesian language, not the Papuan language of the father. This implies a host switch in the language of the man's sons, which will be apparent in the Y chromosome clades. Credit: Authors.

However, we did not immediately pursue alternative explanations. Instead, in Chapter 4 we picked up a different thread, the association of languages with descent groups. Now, with the results of both investigations in hand, we can revisit the as-yet-unexplained phenomena of genetic variation at the scale of islands and villages. How do these pieces—sex, marriage, colonization, language, and clan identity—fit together? Rather than trying to integrate them within a unified Braudellian model of the *longue durée*, we will use phase portraits to investigate specific dynamical relationships, and compare the results with village-level data from the islands of Indonesia.

Sex and power, revisited

In the course of our wanderings among those large and luxuriant islands, we collected kinship, linguistic, and genetic data from many villages. In Chapter 2, we compared the effective population size of male and female ancestors in villages on numerous islands (Table 2.1). And in Chapter 4, we modeled the effects of marriage rules on genetic structure in villages on the islands of Sumba and Timor. In search of historical patterns, in Figure 8.2, we replot these data for all of the sampled villages of Sumba and Timor.

Each dot shows the genetic ancestry of sampled men in a single village. Squares are from patrilineal Sumba; circles are from the mostly matrilineal villages of the Wehali region of central Timor. The axes show the amount of variance, or heterozygosity, on the Y chromosome (inherited from the father) and mtDNA (inherited from the mother) for the men we sampled in these villages. In patrilineal Sumba, men tend to remain in their natal village and seek wives from further afield. This results in a bias: there is more genetic variation (heterozygosity) among the women than the men. The reverse is true in matrilineal Timor: there is greater heterozygosity among the husbands and fathers, who migrate into the village.

Note that these patterns become apparent by analyzing the data at the scale of villages. Why are some villages much more skewed than others? And why are the skewed distributions of villages from the two islands so symmetrical, almost mirror images? This suggests that the underlying cause of the skew is identical in Timor and Sumba, and that the effects are progressive and opposite.

Note further that the skew might also be explained by a different cause: a Darwinian process in which some individuals become socially dominant and produce more children than most people in their village. If this tendency were strong and heritable, it could also cause the observed skew. For example, if a matrilineal village in Timor contained an upper class of ruling Amazons who outreproduced other village women, over time the mtDNA heterozygosity would be reduced, but Y chromosome heterozygosity would not. But for this to explain the entire distribution of villages in Figure 8.2, dominance must be as prevalent among women in matrilineal Timor as for men in patrilineal Sumba. To our knowledge, female dominance of this magnitude has never been observed in any human society.

There is also a third possible explanation for the distribution: Bellwood's *founder-focused ideology*. According to this model, junior members of dominant lineages collect followers and venture forth to found

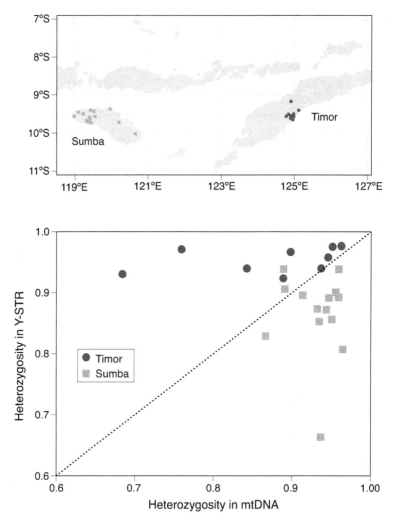

Figure 8.2. Heterozygosity. Levels of diversity on the mitochondrial DNA and Y chromosome arc inversely related on the islands of Sumba (squares) and Timor (circles). Credit: Ning Ning Chung.

new communities. But it is not immediately obvious that this process could produce the surprising variation in ancestries we see in Figure 8.2. Thus, we have three competing explanations for Figure 8.2: marriage rules that favor migration into villages by only one sex, higher reproductive rates by dominant classes, or a founder-focused ideology driving the serial creation of new villages. Call these models *migration, dominance,*

Table 8.1

Variables for the island model and the founder model. Variables in the first section are common for all four cases. Variables in the second section apply only to both the island model and the founder model with dominance. Variables in the third section pertain only to the founder model.

Variable	Symbol	Value
Population size	N	300
Number of villages	V	50
Sampling size	n_s	14–69
Mutation rate (mtDNA)	μ_{mt}	0.0186
Mutation rate (Timor Y-STR)	μ_Y	0.0290
Mutation rate (Sumba Y-STR)	μ_Y	0.0249
Migration rate (male, matrilocal)	m_{mm}	13.5%
Migration rate (female, matrilocal)	m_{fm}	4.5%
Migration rate (male, patrilocal)	m_{mp}	0.5%
Migration rate (female, patrilocal)	m_{fp}	5.5%
Fraction of dominant individuals	δ	0.06
Selective advantage	σ	0.8
Nonheritable reproduction	p	0.1
Population growth rate for founder model	r	0.2
Budding size	N_{bud}	250

and *founder-focused colonization*. They are not mutually exclusive. How can they be evaluated?

Parsing sex, power, and colonization

We pause to consider the nature of the question being asked. The goal is to assess the respective roles of founder-focused colonization, kinship rules related to migration, and social dominance in the independent histories of dozens of villages in the islands of Timor and Sumba. From an analytical perspective this is the most ambitious question we have yet posed, and brings us closest to the questions of interest to anthropologists and prehistorians. Our approach is to proceed sequentially, modeling these processes individually and then in combination. For this purpose we simulate the growth of villages with a stable population size of 300, and sample 50 individuals from each village. These numbers are close to the real averages for both Timor and Sumba. We vary the parameters for migration, dominance, and colonization, and observe the fit with the empirical data.

We begin with migration. We simulated migration (husbands into matrilocal villages or brides into patrilocal communities) at a low rate

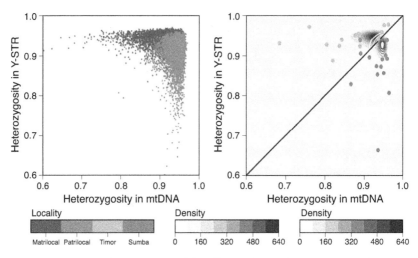

Figure 8.3. Results of simulations of the effects of marital migrations on heterozygosity. Left panel: simulated data. Right panel: comparison of empirical data with simulated data. Females in matrilocal villages and males in patrilocal villages are not migrating. For the dispersing groups, the migration rate is 0.02. 50 villages with a population size of $N = 300$ are simulated 200 times. The mutation rate is set to 0.0249. 40 villagers are sampled from each village. Credit: Ning Ning Chung.

of 2% per generation (Figure 8.3). Because the fit to the data is poor, we increased the migration rate to 10%, which slightly improves the fit (Figure 8.4). Figures 8.5 and 8.6 retain this biased migration and adds dominance. Multiple simulations at varying parameter values show that a combination of relatively high migration plus moderate dominance provides the best fit to the empirical data.

The next step is to consider the effects of colonization: the formation of new villages. In the previous section we defined archaeologist Peter Bellwood's founder-focused colonization model. To assess its observable effects on heterozygosity we need an alternative null model. For that purpose we simulate an "island" model, in which marriages occur between existing villages, and there is no colonization process. In contrast, the founder-focused colonization model simulates a process in which villages grow to a maximum size ($N = 250$), whereupon 50 randomly chosen individuals leave to create a new village. These colonies continue to grow and bud. The results of the founder and island models are compared in Figure 8.7 and Figure 8.8.

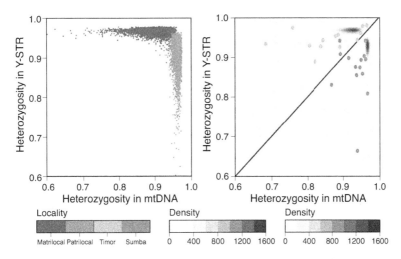

Figure 8.4. Effects of increasing the migration rate to 0.1. Credit: Ning Ning Chung.

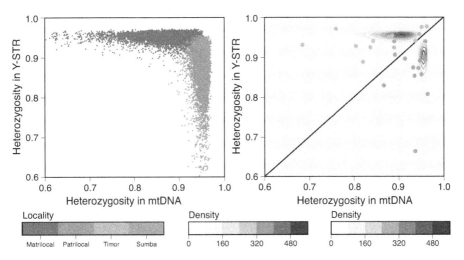

Figure 8.5. Effects of adding dominance to the model. Females in matrilocal villages and males in patrilocal villages migrate at a rate of 0.005. The migration rate of the dispersing groups is 0.04. Dominance parameters used are: $p = 0.1$, $\sigma = 1$, and $\delta = 0.06$. Credit: Ning Ning Chung.

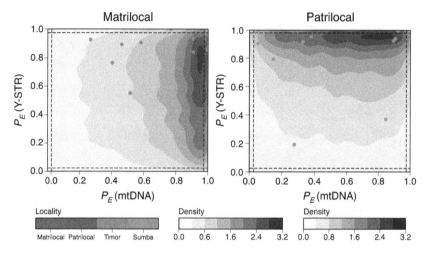

Figure 8.6. Results of Slatkin test on the observed heterozygosity from the above simulation (migration rate = 0.04; $p = 0.1$, $\sigma = 1$, and $\delta = 0.06$). Slatkin tests for each village are shown as dots. (Note: Slatkin's exact test calculates the probability of each possible distribution, and bins them according to whether the probability is greater or smaller than that of the observed distribution. See Chapter 3). Credit: Ning Ning Chung.

The cumulative results can be summarized as follows:

1. The island model without dominance cannot generate the observed heterozygosity or Slatkin distributions.
2. The island model with dominance cannot generate the observed Slatkin distribution.
3. The founder model without dominance cannot generate the observed Slatkin distribution.
4. The founder model with dominance can generate both the observed heterozygosity and Slatkin distribution.

We conclude that the founder model is not only possible, but necessary to explain the observed genetic and language distributions. Our earlier house model correctly predicts that houses will form alliances based on kinship rules, which can be either matri- or patrilineal. But something more is required to explain the observed skew in Sumba and Timor. Bellwood's founder model supplies the missing dynamical process: a persistent founder-focused ideology propelling the formation of new communities. This validates the "cultural core" idea of the Leiden School as clarified by James Fox. It is also consistent with abundant ethnographic evidence for the continuing importance of the founder ideology across

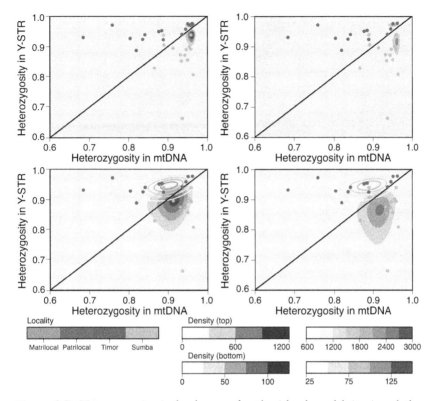

Figure 8.7. Heterozygosity in haplotypes for the island model (top) and the founder model (bottom). Figures on the left are without dominance, while those on the right include the effects of dominance. Parameters used are shown in Table 8.1. Credit: Ning Ning Chung.

the archipelago. Nonetheless, we were surprised by this result. What could account for the persistence of this ideology across the entire Malay archipelago, long after the age of Austronesian colonization came to an end?

Reflections on stone boats

The Tanimbar Islands are located at the far eastern edge of the archipelago. The inhabitants speak Austronesian languages and are organized into houses. As documented by anthropologist Susan McKinnon, the symbolic centers of their villages are stone boats (Figure 8.9), in which the people gather on ritual occasions. The rituals performed in

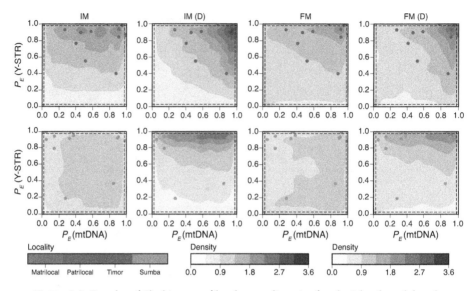

Figure 8.8. Results of Slatkin test of haplotype diversity for the island model and the founder model, with and without dominance. Credit: Ning Ning Chung.

the boats celebrate the real voyages in wooden boats that sustain marital alliances with other houses.[30] Perched on hilltops overlooking the sea, the stone boats charmingly remind us of a point that can easily be lost if we become too absorbed in extracting patterns from the genetic data: houses and kinship systems are based not on biology, but on coherent sets of ideas about society that must be continually renewed. Arguing this point in *Structural Anthropology*, Lévi-Strauss observed that "[n]o one asks how kinship systems, regarded as synchronic wholes, could be the arbitrary product of a convergence of several heterogeneous institutions...yet nevertheless function with some sort of regularity and effectiveness."[31]

What, then, accounts for the coherence and persistence of such "synchronic wholes"? Lévi-Strauss is surely right that the Austronesian house represents more than a strategy for creating alliances. Instead, as Fox argued, it is better understood as the material realization of a set of ideas about the world, based on what Bellwood describes as a founder-focused

[30] S. McKinnon. *From a Shattered Sun: Hierarchy, Gender, and Alliance in the Tanimbar Islands.* University of Wisconsin Press, 1991.
[31] C. Lévi-Strauss, *Structural Anthropology*, Doubleday, 1967 (1963), p. 35.

Figure 8.9. Stone boat in the village of Sangliat Dol, East Yamdena, Tanimbar Islands. Credit: J. Stephen Lansing.

ideology.[32] To explain the extraordinary persistence of the house concept across the archipelago, where nearly everything else was subject to drift, we are drawn to Hegel's argument in the *Phenomenology* about the relationship between ideas and social forms. Hegel's insight is that social institutions have a tendency to become expressed in pervasive metaphors that reflect a society's self-awareness. For this reason, Hegel pays particular attention to cultural products like art, architecture, and mythology that can hold up a mirror to society. In social institutions and religious

[32] Earlier we noted that the key ideas defining this ideology are encoded in metaphors of growth, precedence, and masculine and feminine power. On investigation, we found that these metaphors are more resistant to change than the words associated with kinship systems included in our comparative word lists, which turned out to be as prone to drift as other words. The possibility that this synchronic whole could be retained and transmitted long enough to span the distances between islands and colonization events is demonstrated by our cophylogenetic models of languages and clades.

rites, this mirrored image becomes actualized, as it is in the stone boats.[33] In this sense, social institutions reflect the collective consciousness of a society at a particular historical moment.[34]

Two further points in Hegel's analysis are relevant. First, he argues that societies inevitably seek to impose their mirrored images on the outer world: an expressive process motivated by the desire of reason to make the world congruent with itself. For Hegel, history emerges from the histories of particular peoples within which *Geist* (which can be translated as either "mind" or "spirit," as in the phrase "spirit of the age") assumes some "particular principle on the lines of which it must run through a development of its consciousness and its actuality."[35] Second, Hegel argues that these synchronic wholes are inescapable and change only at the beginning of a new historical epoch.

From this perspective, the house model is more than a kinship system, or even an ideology. Instead, it exemplifies what Hegel meant by a "synchronic whole." Austronesian houses are ritual centers, sites where living generations interact with their gods and ancestors, celebrate marriages and other rites of passage, enact ritual exchanges, and retell the stories of clan origins and myths. We find them everywhere in the archipelago, from the far west (Nias) to the far east (Tanimbar). And we suggest that to explain their historical persistence and coherence, Hegelian as well as demographic insights are required.

Interestingly, however, we do not encounter house societies everywhere in the islands; they are conspicuously absent on Java and Bali. A Hegelian explanation for their absence on those islands would require an account of the origins of a different historical epoch. Can one be discovered?

---

[33] Hegel calls this "objectified reason." In his *Philosophy of History*, Hegel argued that a society's creative accomplishments—its music, philosophy, architecture, mode of governance, mathematics—inevitably build on the accomplishments of the past. These ideas are expressed in works of art, myth, and philosophy. According to Hegel, from one age to the next they acquire greater coherence, drawing ever nearer to ultimate truth. There is also the implication that these ideas aspire to the status of universality, that they represent the culmination of a historical process through which society discovers its own meaning. G.W.F. Hegel, *Vorlesungen über die Philosophie der Weltgeschichte* (Lectures on the Philosophy of History), Cambridge University Press, 1975 (1837).

[34] "Mind is actual only as that which it knows itself to be, and the state, as the mind of a nation, is both the law permeating all relationships within the state and also at the same time the manners and consciousness of its citizens. It follows, therefore, that the constitution of any given nation depends in general on the character and development of its self-consciousness." G.W.F. Hegel, *Phänomenologie des Geistes* (Philosophy of Mind), Oxford University Press, 2018 (1807).

[35] Ibid., paragraph 548.

Emergence of a new synchronic whole in Bali

In the second half of the book, we shifted our gaze from the cultures of the tribal world to the rice-growing villages of Bali. Historical explanations for the uniqueness of Java and Bali invariably focus on the development of irrigation and the appearance of kingdoms, marking the advent of a new historical epoch. But while these developments occurred on both islands at about the same time (the late first millennium AD), afterwards they took different paths, as we saw in chapter 5. The archaeological record suggests that before the advent of irrigated rice, the social organization of these islands was also based on the house model.[36] On Java, house societies were eventually replaced by kingdoms, which resemble those that appeared at the same time in mainland Southeast Asia. Within a few centuries, Javanese houses became peasant villages, no longer linked by marital alliances with each other. Instead, they became individually subject to their rulers.

But in Bali, events took a different course. The earliest Balinese kingdoms closely resembled the first Javanese kingdoms. However, as irrigation systems proliferated on Bali's volcanoes, Balinese royal inscriptions record the emergence of both subaks and water temple networks. These institutions created new forms of cooperative associations between communities. This had two consequences. First, kinship alliances between houses disappeared, and villages ceased to be organized as house societies. Second, as the temple networks grew, the administrative apparatus of kingship gradually disappeared from the royal inscriptions. By the fourteenth century, when the kingdoms of Java were entering an imperial phase, the kingdoms of Bali had disintegrated into feudalities. Both houses and kings gradually disappeared from the inscriptions as a new system of polycentric governance emerged in Bali.

This story has been told elsewhere (in *Perfect Order*), and we will not repeat it here. Instead, in keeping with the theme of this concluding chapter, we return to the subject of chance. Across the archipelago, we find innumerable examples of house societies, but only one society like Bali, where intricate cooperative networks emerged that extended far beyond the scale of local communities. Why only in Bali?

---

[36] The first written records from Java and Bali are inscriptions from the earliest kingdoms. These inscriptions are addressed to communities described by the Proto-Austronesian word *wanua*, which is still in use on many islands, and which we translate as "house." For a summary, see J. S. Lansing. *Perfect Order: Recognizing Complexity in Bali*. Princeton University Press, 2006, pp. 42–54.

Once again, and for the last time, we turn to the phase portraits, in particular the model of adaptive self-organized criticality from Chapter 6. That phase portrait shows that when the trade-off between minimizing losses from pests and water shortages is closely balanced, a phase transition will occur in which cooperation spreads as local and global dynamics become coupled.

From the standpoint of nonlinear dynamics, phase transitions are quite rare; only a few varieties have been discovered. However, they apply to a wide range of phenomena, such as sand-piles, magnetism, and boiling water. These physical systems display universality in a scaling limit, when a large number of interacting parts come together. As the phase transition approaches, the scale-dependent parameters of the physical system diminish in importance, and the scale-invariant parameters dominate. But adaptation plays no role in models of phase transitions in physical systems. Instead, they are triggered by a change in an external control parameter like temperature. Lowering the temperature causes water to transition from gas to liquid, and iron to become magnetized in the presence of a magnetic field. So time is reversible for phase transitions in physical systems, but not in our model, where the control parameter is the local adaptations of the agents (farmers or subaks). When the relative strength of the two parameters, water and pests, are in balance, a phase transition occurs and the local and global dynamics become connected. Details like the liquidity of water or the virulence of pests do not matter, only the nature and strength of their interaction.

But imagine, as a thought experiment, that the phase transitions which linked the subaks did not occur. After all, our analysis shows that these local-to-global transitions can only occur if the environmental parameters are nicely balanced. If as a consequence the interactions between subaks remained exclusively local, would water temple networks exist? Or might Balinese society have developed into a congeries of peasant villages ruled by kings, as happened in Java? We offer this thought experiment in the spirit of Theodor Adorno's reply to Karl Popper, with which we concluded chapter 1: *only through what it is not will it disclose itself as it is.* Still, the Balinese did not abandon the founder-focused ideology of the Austronesians. As in the tribal world, each Balinese clan celebrates and reveres its own unique ancestors. Of equal or greater importance, however, are the gods and goddesses enshrined in the temple networks, who represent the origins of water and life itself. In this way the metaphors of origins and progenitors was extended, indeed universalized in the Hegelian sense, to encompass all the clans.

Afterword

Following this deep dive into the remote past of tribal societies, we antic-ipate that while charitable readers may appreciate both the methods and the stories, they might question their broad relevance for social science. If the most important questions in social science fall within a narrow window—the present and immediate future of modern societies—then our islands of order may seem of limited relevance. Anticipating this reac-tion, we would like to respond with two more stories. Both concern a past that is even more remote than any we have considered so far. Nonetheless, we suggest that they offer fresh perspectives on two of the most urgent questions in social science today.

When we were in graduate school, we were taught that modern humans are the end result of a line of progressive unilineal evolution. Instead it now appears that the genus *Homo* continuously explored many different ways of being human. Hominin species not only diverged and prolifer-ated, sometimes they interbred, and it appears that this was especially true of our own immediate ancestors. Indeed it may be one of their defin-ing characteristics. If so, this could be an argument for broadening our perspective on the evolutionary foundations of human social behavior.

The first story is the discovery of two unknown human ancestors by two postdocs on our team, Guy Jacobs and Georgi Hudjashov, as this book was going to press. Traditionally, human ancestors make themselves known to archaeologists with their fossilized bones. But these ancestors were instead excavated from the genes they left in their human descen-dants. So far, no fossils have been discovered. But if and when they are, and if they contain DNA that can be analyzed, then thanks to Guy and Georgi we can predict in advance the contents of much of their DNA, including several hundred genes for each of the two unknown species. Even without fossilized bones from an archaeological site, we can draw inferences about the relationships of our own ancestors to these close cousins.

The analytical methods that we used to make this discovery do not have direct relevance to complexity science, and we will not provide a detailed explanation here. But the broad outline is easily sketched. At the time of writing, geneticists have sequenced complete genomes for two archaic hominins: Neanderthals and Denisovans. Neanderthals are known from hundreds of fossils, but the existence of the Denisovans is known from DNA that was extracted from nothing more than a single finger bone, which was discovered in 2008 in a cave named Denisova in the Altai mountains of Siberia. Two years later, fossilized teeth from the cave were found to contain DNA sequences like those of the finger bone. The teeth (molars) are much larger than those of any other specimens of the genus *Homo*. So far, no other fossils or tools of Denisovan origin have been found; or if they have, they are not recognized as such. But the finger bone proved to contain enough high-quality DNA to enable geneticists to sequence the entire Denisovan genome. Suddenly it became possible to compare whole genomes for three closely related species: humans, Neanderthals, and Denisovans.

To make such comparisons is technically challenging but conceptually simple: line up the three reference sequences, each consisting of strings of about four billion DNA nucleotides, and see where they differ. Then, use molecular clocks to estimate when these species shared a common ancestor. As of now, the earliest anatomically modern humans date to approximately 198,000 years ago. The separation of Denisovan and Neanderthal clades is estimated to have occurred between 470,000 and 380,000 years ago. The separation of both of these archaic groups and the direct ancestors of modern humans is much older, about 770,000 to 550,000 years ago. Consequently, archaic Neanderthals and Denisovans were cousins, more closely related to each other than to our human ancestors. But now the story gets interesting. Genes from Neanderthals and Denisovans have been detected in modern human populations. Geneticists infer that our ancestors sometimes mated with these sister species and produced children who were absorbed into modern human populations. These social interactions can be approximately dated as beginning around 50,000 years ago.

In 2018 we obtained whole genome sequences for 161 of our genetic samples, which we selected from 14 islands spanning Island Southeast Asia from Sumatra to New Guinea. Jacobs and Hudjashov lined up these complete genomes and compared each of them with the three reference sequences—humans, Neanderthals, and Denisovans. They discovered that modern Papuans carry hundreds of Denisovan genes, and these genes came from two different Denisovan groups, which differed in important ways from the Altai reference individual. In other words, there were not one but at least three groups of Denisovans: the one represented by the finger bone from the cave, and two others which are known only by the different sets of genes that are preserved in the DNA of modern Papuans. For simplicity, we designate these groups as D0 (the form decoded from the finger bone), D1, and D2. All three descend from a common Denisovan ancestor, from which D2 split off about 363,000 years ago, and D1 about 283,000 years ago.

This raises a question: why did D1 and D2 mate with humans, but seemingly not with each other? Even infrequent matings would have kept the Denisovans intact as a single species. Instead, in north Asia, D0 mated with the ancestors of East Asians, Siberians, and Tibetans, while in the

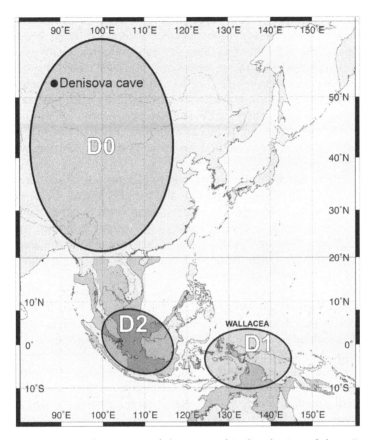

Figure 8.10. A speculative map of the geographic distribution of three Denisovan clades that human ancestors encountered and mated with, based on the distribution of Denisovan genes in modern human populations. Credit: Yves Descatoire.

islands, D1 and D2 mated at different times and places with the ancestors of the Papuans (Figure 8.10).[37]

There is growing evidence that the willingness of our human ancestors to mate with other branches of the hominin family was advantageous. For example, D0, the apparent source of a version of the EPAS1 gene that regulates the body's production of hemoglobin, is abundant in Tibetans and

[37] G. S. Jacobs, G. Hudjashov, L. Saag, P. Kusuma, C. C. Darusallam, D. J. Lawson, M. Mondal, L. Pagani, F.-X. Ricaut, M. Stoneking, M. Metspalu, H. Sudoyo, J. S. Lansing, and M. P. Cox. 2019. Multiple deeply divergent Denisovan ancestries in Papuans. *Cell* 77:1010–21.

provides an advantage in the oxygen-starved heights of the Himalayas.[38] It appears that the three Denisovan clades occupied different geographical ranges at the different times that migrating humans encountered them. For example, peoples on the mainland of Papua have more D1 genes than the Baining people on the island of New Britain, but similar numbers of D2 genes. New Britain was settled by humans at least 35,000 years ago. The implication is that archaic Denisovans were present in Papua and its environs before the arrival of modern humans, and that the ancestors of mainland Papuans interacted in different ways than the Baining with the D1 Denisovan group.

We suggest that these discoveries have implications for our understanding of the evolution of human biology and social behavior. We now know that our immediate ancestors were strongly inclined to explore, migrate, and hybridize. These tendencies are unlikely to appear in a list of social attributes based on studies of contemporary xenophobic societies. But perhaps they have fallen away since the Pleistocene and are no longer relevant? This question brings us to our final story.

Borneo is the largest island in Indonesia, and contains the greatest terrestrial biodiversity on our planet. So far, there has been little genetic research on the peoples of Borneo. In the last few decades, more than half of the island's primary forests have been logged and replanted with oil palms. As a consequence of this loss of the forests, the seas around the island contain more floral debris than the Amazon. In the remaining forests, which are mostly in the central and northern uplands, there is a small population of a few thousand hunter-gatherers called Punan. All known Punan groups are no longer full-time hunter-gatherers. They have been resettled in permanent communities and encouraged to become farmers or laborers. In 2018, we began to study Punan communities. Anthropologists have long debated whether the Punan are the descendants of an ancient migration of hunter-gatherers or of Austronesian farmers who gradually abandoned farming in favor of a hunting and gathering way of life. This question should be resolvable using DNA to trace the origins of contemporary Punan communities.

We began our research in Long Sule, an isolated Punan community in northwest Borneo accessible only by missionary aircraft. The village is located on a river in the midst of primary forest, and consists of wooden houses where several formerly nomadic Punan groups settled a generation ago. Today they continue to hunt and gather, but also practice swidden farming and collect *gaharu* (fragrant wood) to sell to traders.

---

[38] E. Huerta-Sanchez, X. Jin, Asan, et al. 2014. Altitude adaptation in Tibetans caused by introgression of Denisovan-like DNA. *Nature* 512:194–7.

Our team of medical researchers from the Eijkman Institute for Molecular Biology moved into one of the houses, and for a week, we ran a medical clinic while gathering information about health, kinship, demography, genetics, and language. We bathed in the pristine river, and the people fed us with fish and meat from a wild boar.

We were interested in two comparative questions about the Punan. The first was their resemblance to the well-known hunter-gatherers of southern and East Africa, the Khoisan and Hadza. These groups have been conscripted by the popular media, as well as anthropologists, as the canonical exemplars of the hunting and gathering way of life. The second question was a comparison of the Long Sule community with conditions on the Punan reservation in the town of Malinau.

We can offer only preliminary observations here on all of these questions. After leaving Long Sule, we carried out three more community studies, one of them in the Punan reservation. As we were organizing the research at the reservation, we were approached by the elected leader of the Punan of Eastern Kalimantan, who informed us that he had recently visited an isolated community of cave-dwelling Punan who are still hunter-gatherers and (perhaps wisely) flee from contact with outsiders. The Punan leader explained that he was concerned about their health and regarded it as his duty to give them whatever help they may feel that they need. To that end, Lansing and the Eijkman team accepted his invitation to visit the cave, which is said to be one of several.

Like the well-known hunter-gatherers of Africa, the Punan of Long Sule divide and share the meat from successful hunts, as they did with us. They also pursue their new occupation as farmers in much the same collectivist spirit: each year, they meet to decide on a stretch of a river where they will work together to create their swidden gardens. The women have their own favorite bathing pools in the river, but they were happy to invite all of us to join them, and gave us an admittedly superficial impression of gender equality. Their diet is varied and their health appears to be better than that of the farmers we have studied on other islands. However, although their forests are still intact and the rivers are clean and full of fish, there are plans for the expansion of logging, oil palms, and mining concessions into the region (Figure 8.11).

Not long ago, the loss of the forests and the resettlement of the last remaining Punan would have merited little more than a footnote for social scientists. We wish to conclude our book by offering a different perspective, directed toward the evolutionary foundations of social behavior and the scope and subject matter of social science. The Punan offer a unique opportunity to broaden our knowledge of the hunting and gathering adaptations that shaped the evolution of all hominins. But contemporary social science has become nearly exclusively concerned with modern

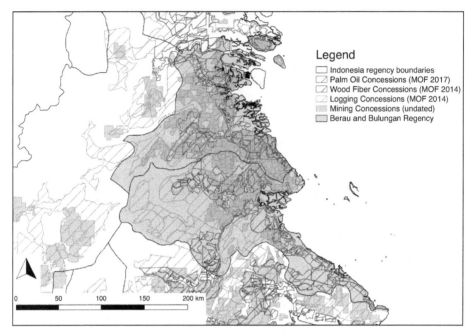

Figure 8.11. Industrial concessions for the Berau and Bulungan Regencies of North East Kalimantan. Credit: Janice Teresa Lee Ser Huay.

consumer societies. We suggest two reasons for broadening that focus. First, a robust European tradition of social theory argues that consumer economies transform subjects into consumers by progressively narrowing both subjective identity and social relationships. Consequently, to sample exclusively from consumer populations restricts the scope of social analysis and may skew the conclusions. Second, the rapid expansion of consumer society since the end of World War II, in what has been called the Great Acceleration,[39] is widely predicted to be unsustainable. Simply put, the Anglo-American tradition of equilibrium social science that we invoked in the introduction to this book looks increasingly dated, especially in the context of evolutionary theory, and is arguably an obstacle to developing a more nuanced and powerful approach to understanding human society and social-environmental conditions in the twenty-first century.

[39] W. Steffen, W. Broadgate, L. Deutsch, et al. 2015. The trajectory of the Anthropocene: The Great Acceleration. *The Anthropocene Review* 2:81–98.

Correlation Functions

Lock Yue Chew

We use a definition of the correlation function $C(d)$ that is based on the mutual information between the cropping pattern X at site i and the cropping pattern Y at site j, where the distance from site i to j is d. The mutual information measures how much the knowledge of the cropping pattern at one site reduces the uncertainty on knowledge of the cropping pattern at the other site. It is defined as

$$C(d) = \frac{1}{\mathcal{N}} \sum_{X=1}^{4} \sum_{Y=1}^{4} P_d(X, Y) \log_2 \frac{P_d(X, Y)}{P_d(X)P_d(Y)} \qquad (A.1)$$

where $P_d(X, Y)$ is the probability of cropping patterns X and Y occurring at sites that are some distance d apart. Note that X and Y take values from 1 to 4 with "1 = green," "2 = red," "3 = blue," and "4 = yellow." Operationally, the joint probability $P_d(X, Y)$ is determined by taking the relative frequency of the cropping patterns X and Y against all possible combination of cropping patterns between sites at a relative distance d. Note that the site here refers either to a pixel in the satellite image or to a lattice site for the model. The marginal probability of cropping pattern X or Y is $P_d(X)$ or $P_d(Y)$. $\mathcal{N}$ is the normalization constant equal to the Shannon entropy of the cropping pattern X; i.e., $\mathcal{N} = -\sum_{X=1}^{4} P_0(X) \log_2 P_0(X)$. This ensures that the correlation is normalized, so that $C(d=0) = 1$. We use this definition for the correlation function because it is applicable to random variables in symbolic form. The standard correlation function in two dimensions is inappropriate, as it needs random variables in numeric form. However, these two definitions for the correlation functions are closely related if the joint probability distribution is Gaussian.[1]

---

[1] A. M. Fraser. 1989. Reconstructing attractors from scalar time series: A comparison of singular system and redundancy criteria. *Physica D: Nonlinear Phenomena* 34:391–404.

The correlation length ε is defined as the variance (second moment) of the correlation function from equation A.1:

$$\varepsilon = \left(\frac{\sum_d d^2 C(d)}{\sum_d C(d)} \right)^{\frac{1}{2}} \tag{A.2}$$

A systematic study was performed of the dependence of the average harvests H, the power law exponents α, and the correlation lengths ε on the parameters ρ and δ. The results are shown in Figure 6.5. Here, we observe the emergence of critical behavior at a region where water and pest stresses balance as adaptation progresses in the simulation. This region is highlighted with white lines in Figure 6.5B. A comparison with the observed data for the power law exponent α in Table 6.1 suggests that model results from this parameter region is compatible with the empirical data. At the critical region, the entire system of farms becomes correlated, as global control emerges through simple local interactions between farmers.

Game and Lattice Models

Here we explain the differences between the two-player game[1] and the lattice model used in the present analysis. We also discuss an important variant of the lattice model. The game in Lansing and Miller describes an idealized situation involving two farmers, in which the upstream farmer can control the flow to the downstream farmer. This puts the downstream farmer at a disadvantage, a problem known as the *tail-ender* problem in irrigation studies.[2] We modeled the tail-ender problem on the lattice by introducing a y-axis. The top of the lattice, $y = L = 100$, is at the water source, the bottom, $y = 0$, is furthest downstream. Water stress for a farm is defined as the fraction of cells at its y-coordinate and higher upstream that follow the same irrigation strategy. In the resulting model of upstream dominance, harvests quickly decline along the downstream gradient and barely change throughout the simulation.

However, the game shows that when pest damage is taken into account as well as water stress, both upstream and downstream farmers have an incentive to cooperate (synchronize irrigation). Thus, for both upstream and downstream farmers, whether or not cooperation is their best strategy depends on the balance between pest and water stress. In the lattice model described in the main text, with no y-axis, the adaptive selection of irrigation schedules by individual farmers equalizes water sharing at the phase transition. The game and the lattice model are not directly comparable, because pest and water stresses are calculated differently. But they offer complementary insights: the game captures the logic of the pest-water trade-off, while the lattice shows how cooperation can spread in a coupled system, where farmers adapt to the pest and water stresses triggered by their own decisions.

In the two-player game, coordinated cropping occurs between the two farmers when $\rho > \delta/2$. In the $L \times L$ lattice, pest stress ρ is computed as $a/(m + f_p)$ and water stress δ as bf_w. The constant m serves to bound the maximum possible stress from pests at $f_p = 0$. In other words, the maximum pest stress is a/m. The variable f_p gives the fraction of neighbors of

[1] J. S. Lansing and J. H. Miller. 2005. Cooperation, games and ecological feedback: Some insights from Bali. *Current Anthropology* 46:328–34.

[2] "Whatever the reason, the tail-end problem is instantly recognizable in the field." A. Laycock. *Irrigation Systems: Design, Planning and Construction.* Cabi, 2011.

a farmer that share the same cropping pattern as within a radius r, while the variable f_w defines the fraction of all lattice sites that have the same cropping pattern as the farmer (see also main text for the definition).

It is important to note that the model is effectively one-dimensional with respect to the relative weight b/a, which explains the appearance of critical behavior along straight lines such as $b/a \approx 20$. The gradient of the line depends on the constant m in the denominator of pest stress. The model results of the correlation lengths are shown in Figure 6.5C, where the white region indicates the empirical range (see also Table 6.1). The parameter region here is slightly narrower, and is also closely centered around $b/a \approx 20$. This critical region separates two phases of the model: one where water stress dominates and the other where pest stress dominates.

As the phase portrait shows, across a wide range of parameter values local adaptation will reduce both pest and water stress, initially in local neighborhoods. At the phase transition, these stresses balance each other while harvests are optimized (Figure 6.5). The equalization of water sharing in the lattice model is not assumed from the start, but emerges at the phase transition within a certain parameter range.[3] The resulting mosaic of correlated irrigation schedules is consistent with the satellite imagery.

---

[3] Irrigation schedules are allocated randomly to farms at the start of the simulation. Subsequently they fluctuate during trial-and-error local adaptations until the model reaches its attractor. They nearly equalize only at the critical transition and quadrant states.

Intra- and Intersubak Coordination of Irrigation

Both the lattice model and the game were designed to be as simple as possible, with the goal of exposing the underlying dynamics of cooperation in the subaks. How well do they succeed?

In reality, Balinese farmers acquire the right to use irrigation water by making offerings or prestations to the Goddess of the Lake(s), who "makes the waters flow." This principle is given physical reality by means of proportional dividers in the canals, which instantiate a fractional division of the water flows in units called *tektek*, which determine the debt owed to the Goddess. For example, each subak in the congregation of the Masceti Pamos Apuh water temple (Figure 6.6) has the right to a share of water so long as they contribute offerings and support (*soewinih*) to the temple proportional to their tektek allocation. Within each subak, rights to a proportional share of irrigation flow are based on smaller proportional dividers, called *tektek alit* (small tektek), which are equal for the entire subak. Farmers who fail to meet their responsibilities for maintenance of the irrigation systems and offerings to the Goddess are at risk of having their water rights terminated by a decision of the subak meeting.

It is easy to verify the fairness of small tektek allocations within subaks by checking the flows at the proportional dividers. Any deviation is an immediate cause for concern, and will soon be corrected by the subak. To discover whether tektek shares between subaks are also equitable, in earlier work[1] we measured irrigation flows at the intakes to the primary canals for ten subaks that jointly coordinate their irrigation schedules at the regional water temple Masceti Pamos Apuh (Table 6.2 and Figure 6.6). The *r* correlation is nearly perfect. In a survey of 150 farmers from these 10 subaks, in answer to the question "Is the division of water by the Pamos water temple equitable?," all said yes.[2]

When cooperation breaks down, subaks cease to monitor the allocation of tektek and tektek alit, and farmers are free to plant whenever they like. This condition is described as *tulak sumur* (translated by

[1] J. S. Lansing, M. P. Cox, S. S. Downey, et al. 2009. A robust budding model of Balinese water temple networks. *World Archaeology* 41:112–33.

[2] J. S. Lansing. *Perfect Order: Recognizing Complexity in Bali*. Princeton University Press, 2006.

Lansing as "reject the wellspring" and in a Balinese-English dictionary less idiomatically as "sow rice at the wrong time"[3]). In tulak sumur, upstream farmers are under no obligation to share water with their neighbors. However, the resulting asynchrony of rice growth in adjacent fields creates ideal conditions for the rapid growth of rice pests, including rats, insects, and insect-borne diseases

Tulak sumur was legally mandated in Bali in the 1970s to support the introduction of the Green Revolution rice, by encouraging farmers to plant as often as possible. This triggered the tail-ender problem, because upstream farmers were able to plant continuously, while downstream farmers often could not plant at all. As noted above, this policy led to a rapid buildup of pest populations and harvest failures, and was soon discontinued. Still, from time to time some subaks abandon the goal of collectively enforcing equal water shares, which leads to tulak sumur.[4] Three examples of such breakdowns at the subak level are discussed in Chapter 4 of Lansing's *Perfect Order*.[5] In all cases, the farmers soon returned to the consensual management of irrigation by their subak.

How well does the simple decision rule in the lattice model (imitate your most successful neighbor) capture the process by which farmers and subaks adjust their irrigation schedules? In reality, the farmer's decisions reflect the imperatives of the terraced landscape, where fields must be kept flat and protected by bunds to turn them into shallow ponds. The average farm is about 0.3 hectares and consists of many small adjacent ponds. Peak irrigation demand occurs at the beginning of each planting cycle, to create the ponds. Afterwards, the tiny irrigation channels that connect the ponds require continuous monitoring. Farmers often borrow water from their upstream neighbors, a debt that can be repaid later on by temporarily blocking the flow to their own fields. A farmer who cannot borrow water from one upstream neighbor can try to borrow from others whose fields are either adjacent to the first upstream neighbor or further upstream. For these reasons, decisions about water sharing, irrigation schedules, and the need for pest control begin with conversations among small groups of neighboring farmers. Importantly, this is true for upstream farmers as well as those whose fields are located downstream. Subak meetings provide a venue to reach a consensus. An analogous process occurs in the lattice model, as neighbors create synchronized patches that eventually become correlated, equalizing water sharing.

---

[3] N. Shadeg, Balinese-English Dictionary. 2007. Tuttle Publishing, p. 458.

[4] J. S. Lansing and T. De Vet. 2012. The functional role of Balinese water temples: A response to critics. *Human Ecology* 40:453–67.

[5] Lansing, *Perfect Order*.

This result sheds new light on the tail-ender problem in the Balinese context. Several authors have argued, largely on *a priori* grounds, that some form of centralized water control by Bali's rulers must have existed in the past. In an earlier publication, we evaluated these claims and argued that no historical or empirical evidence exists in support of this claim.[6] But how then was the tail-ender problem solved? Adaptive self-organized criticality offers an explanation.

[6] Lansing and De Vet, The functional role of Balinese water temples, 453–67.

~~~~~~

Fisher Information

Lock Yue Chew and Ning Ning Chung

Fisher Information, F, is a measure of how much information an observation of a random variable X provides about an unknown parameter θ of a probability distribution $p(x|\theta)$. With the log-likelihood of the parameter θ $L(\theta|x) = \ln((p(x|\theta))$ and assuming a continuous X,

$$F(\theta) = \int_x \left(\frac{dL(\theta|x)}{d\theta}\right)^2 p(x|\theta)dx \tag{D.1}$$

Intuitively, this weights the probability of observations by the extent to which they constrain θ. A flat log-likelihood surface of θ with respect to X implies that an observation provides little information about θ, and the Fisher Information is low; conversely, if the log-likelihood surface is sharply peaked, then it is relatively easy to constrain θ through observations, and the Fisher information is high.

We use Fisher Information to examine the relationship between PCA—which seeks linearly uncorrelated components that capture the greatest variance among all individuals in the survey—and the population units in which these individuals exist. Specifically, we are seeking to identify subaks containing individuals who vary in ways that are not captured by the system-level PCA (Figure 7.4). Such subaks may exhibit different internal dynamics.

Having generated the PCA based on the reduced 19-question set of responses from all individuals, for each subak we:

1. Project all members of that subak onto a principal component axis.
2. Fit three statistical distributions (Gaussian, Rayleigh, or Pareto) to the distribution of individuals on that axis by maximum likelihood and calculate their sum of squared errors (sse). The distribution with the lowest sum of squared errors is accepted as the distribution that best describes the data.
3. Calculate the Fisher Information $F(\mu)$ of that distribution, where μ is the first statistical moment.

This is repeated for the first and second principal components. We then use a Gaussian kernel density estimator to interpolate a Fisher

Information surface onto the PCA (Figure 7.4). The equations of $F(\mu)$ for the three distributions can be derived from Eq. D.1:

$$F(\mu_{\text{Gaussian}}) = \frac{1}{\sigma^2}, \tag{D.2}$$

$$F(\mu_{\text{Rayleigh}}) = \frac{2}{\pi} \int_x \left(\frac{x^2}{s^3} - \frac{2}{s}\right)^2 p(x|s) \, dx, \tag{D.3}$$

$$F(\mu_{\text{Pareto}}) = \frac{(b-1)^2}{b^2 s^2} (3b^2 - 3b + 1), \tag{D.4}$$

with the probability density function (pdf) of the Gaussian distribution given by

$$p(x|\sigma) = \frac{1}{\sqrt{2\pi\sigma^2}} \exp(-\frac{x^2}{2\sigma^2}), \tag{D.5}$$

while the probability density function of the Rayleigh distribution is

$$p(x|s) = \frac{x}{s^2} \exp(-x^2/(2s^2)), \tag{D.6}$$

and the probability density function of the Pareto distribution is

$$p(x|s, b) = \frac{bs^b}{x^{b+1}} \tag{D.7}$$

In general, symmetrical distributions are better fitted by a Gaussian distribution, while skewed distributions are better fitted by a Rayleigh distribution. The Pareto distribution is a better fit for distributions with a sharp peak. To discover which distribution is the best fit for each subak, we test all three distributions.

APPENDIX E

Energy Landscape Analysis

Lock Yue Chew and Ning Ning Chung

The data are normalized and packaged into an $N \times M$ data matrix, where N is the number of farmers who participated in the survey and M is the number of descriptors. We performed principal component analysis on the matrix to yield eigenvalues and eigenvectors of the principal axes. The state of each subak is then projected onto the principal axes ($i = 1, 2, \cdots, M$) to yield $\{d_i^1, d_i^2, \cdots, d_i^S\}$, where S is the total number of subaks studied. We consider only the first three components of the projected states; i.e., $\{d_i^1, d_i^2, \cdots, d_i^S\}$, with $i = 1, 2$, and 3, and ignore the rest. We then follow the energy landscape analysis to binarize these states by forming a sequence of binarized components $\{\sigma_i^1, \cdots, \sigma_i^S\}$. If d_i^s is greater than a threshold, $\sigma_i^s = 1$; otherwise, $\sigma_i^s = -1$. The threshold is arbitrary, and we set it such that the conditional probability $P(\sigma_i^{s_1} = \sigma_i^{s_2} \mid |d_i^{s_1} - d_i^{s_2}| \leq \epsilon)$ is maximized for all i, with $\epsilon = 0^+$. Thus, the state of a subak is given by a three-dimensional vector $\sigma = (\sigma_1, \sigma_2, \sigma_3) \in \{-1, 1\}^3$, where we have suppressed s. Note that there are 2^3 possible states in total.

From the states of the S subaks, we compute the relative frequency with which each state is visited, $P_{\text{empirical}}(\sigma)$ (Table E.1). We then fit the distribution to a Boltzmann distribution given by Ezaki and colleagues[1]

$$P(\sigma \mid h, J) = \frac{\exp[-E(\sigma \mid h, J)]}{\sum_{\sigma'} \exp[-E(\sigma' \mid h, J)]}, \tag{E.1}$$

where

$$E(\sigma \mid h, J) = -\sum_{i=1}^{3} h_i \sigma_i - \frac{1}{2} \sum_{i=1}^{3} \sum_{j \neq i} J_{ij} \sigma_i \sigma_j \tag{E.2}$$

is the energy and $h = \{h_i\}$ and $J = \{J_{ij}\}$ ($i, j = 1, 2$, and 3) are the parameters of the model. We assume $J_{ij} = J_{ji}$ and $J_{ii} = 0$ for ($i, j = 1, 2$, and 3).

According to the principle of maximum entropy, we select h and J such that $\langle \sigma_i \rangle_{\text{empirical}} = \langle \sigma_i \rangle_{\text{model}}$ and $\langle \sigma_i \sigma_j \rangle_{\text{empirical}} = \langle \sigma_i \sigma_j \rangle_{\text{model}}$, where

[1] T. Ezaki, T. Watanabe, M. Ohzeki, and N. Masuda. 2017. Energy landscape analysis of neuroimaging data. *Philosophical Transactions of the Royal Society A* 375:0287.

Table E.1
Frequencies and energies for the 2^3 binarized states.

State	1	2	3	4	5	6	7	8
σ_1	-1	-1	-1	-1	1	1	1	1
σ_2	-1	-1	1	1	-1	-1	1	1
σ_3	-1	1	-1	1	-1	1	-1	1
$P_{empirical}(\sigma)$	0.65	0.15	0.05	0.05	0.05	0.05	0.00	0.00
$P_{model}(\sigma)$	0.6487	0.1504	0.0505	0.0487	0.0505	0.0487	0.0005	0.0021
$E(\sigma)$	-3.07	-1.61	-0.52	-0.48	-0.52	-0.48	4.04	2.65

$\langle \cdots \rangle_{empirical}$ and $\langle \cdots \rangle_{model}$ represent the mean with respect to the empirical distribution and the model distribution, respectively. We use a likelihood maximization algorithm to estimate the parameters of the model; i.e., h and J.

With h and J, we obtain energies of all states and construct a dendrogram using the following procedure. We enumerate local minima; i.e., the state whose energy is smaller than that of all neighbors. Here, we define neighbouring states σ and σ' as states that are only different at the third principal axis. For example, $(-1, -1, -1)$ and $(-1, -1, 1)$ are nearest neighbor states. We consider distance along the third principal axis to be the smallest because the third principal component is the least significant of the first three components in PCA. Based on this definition, the nearest neighbor states are states 1 and 2, states 3 and 4, states 5 and 6, and states 7 and 8 (see Table E.1). A local minimum would reside within one of these nearest neighbor states. Thus, a connection from one energy minimum to another energy minimum has to occur between states that differ in at least the first two principal axes. Such a connection signifies a branch point of the dendogram and represents an energy barrier between the two energy minima.

Each local minimum has a basin of attraction in the state space, with each state belonging to one of the attraction basins. By repeatedly following a neighboring state that has the smallest energy value, the associated local minimum can be reached. For a given pair of local minima α and α', we consider a path connecting them as a transition path. The path connecting them includes a sequence of states that begin at state α and end at state α'. The largest energy value among the states on the path gives the energy barrier that needs to be overcome for the transition to happen. With the information of all attractors and energy barriers between them, we construct a hypothetical two-dimensional landscape.

Survey Questions Used in Chapter 7

A questionnaire comprising 52 questions was used in a survey of farmers in 20 subaks by Lansing in 2013. Because not all questions were relevant for the transition paths analysis, a set of 35 questions were used in the preliminary analysis (Table F.1). In an earlier study,[1] we performed a higher-order clustering and reduced the number of descriptors from 35 to 19. Descriptors that are relatively insignificant were removed. The subset of 19 questions was used in the more detailed analysis (highlighted questions, see Table F.1): 2, 3, 12, 13, 14, 16, 17, 18, 21, 22, 23, 24, 26, 27, 29, 30, 32, 33, 34. In Figure F.1, we show the cluster map for the correlation matrix.

Translation of survey questions for chapter 7

Note: Sawah are wet rice terraced fields. Encoding of responses is indicated in square brackets following answer options.

1. Name of subak
2. Number of subak members
3. Where is your largest farm?
    (a) same village where I live
    (b) different village but same subdistrict as my residence
    (c) different subdistrict, same district (kabupaten)
4. Amount of sawah you farm now:
    (a) Owned by me _____hectares
    (b) I sharecrop _____hectares
5. Have you inherited sawah?
    (a) yes [1]
    (b) no [0]
6. Have you purchased sawah?
    (a) yes [1]
    (b) no [0]

[1] H. S. Sugiarto, J. S. Lansing, N. N. Chung, et al. 2017. Social cooperation and disharmony in communities mediated through common pool resource exploitation. *Physical Review Letters* 118:208301.

Table F.1

A set of 35 questions, obtained from the survey comprising 52 questions (see below), were used in the preliminary analysis. A subset of 19 questions (highlighted) were used in the more detailed analysis.

Number	Descriptor	Question Number
1	Own farmland	Q4a
2	Sharecrop land	Q4b
3	Inherit a farm	Q5
4	Purchase	Q6
5	Sold farm	Q7
6	Income	Q8
7	Harvest	Q9
8	Satisfaction with harvest	Q10
9	Origin	Q11
10	Condition of canals	Q12
11	Condition of fields	Q13
12	Synchronize	Q16
13	Attendance at meetings	Q17
14	Participation in maintenance	Q18
15	Attendance at rituals	Q19
16	Accept subak decisions	Q20
17	Water shortages in subak	Q21
18	Water shortages myself	Q22
19	Pests damage in subak	Q23
20	Pests damage myself	Q24
21	Thefts of water	Q25
22	Conflicts among members	Q26
23	Choice of subak head	Q27
24	Fines	Q28
25	Crop schedule followed	Q29
26	Plan work	Q30
27	Written rules followed	Q31
28	Fines frequency	Q36
29	Condition of subak	Q40
30	Decisions of subak accepted	Q41
31	Technical problems	Q42
32	Social problems	Q43
33	Caste problems	Q44
34	Class problems	Q45
35	Resilience	Q46

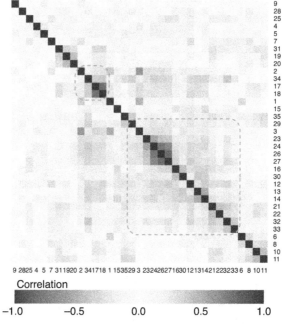

Figure F.1. Cluster map for the correlation matrix of the 35 descriptors. The reduced set of the 19 descriptors are marked in dashed boxes. Credit: Authors.

7. Have you sold sawah?
   (a) yes [1]
   (b) no [0]
8. Your income from sawah, compared to your total family income, is:
   (a) 100% [1.0]
   (b) 66% [0.66]
   (c) 50% [0.5]
   (d) 25% [0.25]
   (e) 10% [0.1]
9. Your last harvest of beras (unhulled rice)_____kg/ha. When? Which rice variety?
10. Your level of satisfaction with this last harvest?
    (a) very satisfied [4]
    (b) satisfied [3]
    (c) less satisfied [2]
    (d) not satisfied [1]

11. Were you born in this village?
    (a) yes [1]
    (b) no [0]
12. The condition of your subak's irrigation canals now, compared to the usual in Bali:
    (a) very good [4]
    (b) good [3]
    (c) average [2]
    (d) not good [1]
13. The condition of your subak's sawah now, compared to the usual in Bali:
    (a) very good [4]
    (b) good [3]
    (c) average [2]
    (d) not good [1]
14. Which is usually worse, irrigation water shortage or pest damage?
    (a) water shortage
    (b) pests
    (these were not converted to a numerical scale)
15. Where is your sawah located?
    (a) upstream sector of my subak
    (b) midsection
    (c) downstream sector of my subak
    (these were not converted to a numerical scale)
16. How many subak members follow (a decision to) plant at the same time?
    (a) all 100% [1.0]
    (b) most [0.85]
    (c) half [0.5]
    (d) less than half [0.4]
17. How many subak members (regularly) attend every subak meeting?
    (a) more than 80% [0.85]
    (b) between 60% and 80% [0.70]
    (c) less than 60% [0.5]
18. How many subak members regularly participate in each collective work activity such as working on the canals?
    (a) more than 80% [0.85]
    (b) between 60% and 80% [0.70]
    (c) less than 60% [0.5]

19. How many subak members regularly participate in religious activities carried out by the subak, like offerings?
    (a) more than 80% [0.85]
    (b) between 60% and 80% [0.70]
    (c) less than 60% [0.5]
20. How many subak members usually follow (obey) the decisions of the subak head and the subak meetings?
    (a) all [1]
    (b) most [0.75]
    (c) half [0.50]
    (d) less than half [0.25]
21. Are water shortages frequent in the subak during the dry season?
    (a) never [4]
    (b) seldom [3]
    (c) sometimes [2]
    (d) frequent [1]
22. Do you yourself frequently experience water shortages?
    (a) never [4]
    (b) seldom [3]
    (c) sometimes [2]
    (d) frequent [1]
23. Are there frequent pest infestations?
    (a) never [4]
    (b) seldom [3]
    (c) sometimes [2]
    (d) frequent [1]
24. Do you yourself frequently experience pests?
    (a) never [4]
    (b) seldom [3]
    (c) sometimes [2]
    (d) frequent [1]
25. Is water theft frequent in the subak?
    (a) never [4]
    (b) seldom [3]
    (c) sometimes [2]
    (d) frequent [1]
26. Are (social/personal) conflicts frequent among subak members?
    (a) never [4]
    (b) seldom [3]
    (c) sometimes [2]
    (d) frequent [1]

Concerning decisions or results from subak meetings, which of the following reflect real democracy and which are just a democratic veneer?

27. Selection of subak head
    (a) veneer of democracy [0]
    (b) democratic/just (i.e., fair) [1]
28. Fines
    (a) veneer of democracy [0]
    (b) democratic/just (i.e., fair) [1]
29. Choice of cropping pattern
    (a) veneer of democracy [0]
    (b) democratic/just (i.e., fair) [1]
30. Organization of collective work
    (a) veneer of democracy [0]
    (b) democratic/just (i.e., fair) [1]
31. Reading and following written rules of the subak
    (a) veneer of democracy [0]
    (b) democratic/just (i.e., fair) [1]
32. If a farmer sees a person stealing water, what's the best thing to do?
    (a) ignore it and have faith in karmic consequences
    (b) speak to the person
    (c) report to the subak head
    (d) bring it up in the subak meeting
    (e) it depends...
    (these were not converted to a numerical scale)
33. Suppose a subak head collected money for irrigation repair, and after the work was done there was still money left unspent. The subak head kept some but not all of the money. If this became known, what would be the best response?
    (a) ignore it and have faith in karmic consequences
    (b) speak to the person
    (c) report to the subak head
    (d) bring it up in the subak meeting
    (e) it depends...
    (these were not converted to a numerical scale)
34. Who has the right to inherit sawah here?
    (a) all sons equally
    (b) the youngest son
    (c) the oldest son
    (d) other
    (these were not converted to a numerical scale)

35. Who has the right to decide about imposing fines?
    (a) subak head
    (b) subak meeting
    (c) both
    (d) neither has the right
    (these were not converted to a numerical scale)
36. Are fines frequently imposed in your subak?
    (a) never [0.01]
    (b) seldom [0.5]
    (c) often [1]
37. Are you married?
    (a) yes
    (b) no
    (these were not converted to a numerical scale)
    If respondent is married:
38. Was your wife born in this community (this subak)?
    (a) yes
    (b) no
    (these were not converted to a numerical scale)
39. Is your wife a first or second cousin?
    (a) yes
    (b) no
    (these were not converted to a numerical scale)
40. The condition of your subak now is:
    (a) excellent, still intact [5]
    (b) good enough [4]
    (c) some problems have begun [3]
    (d) not good [2]
    (e) bad [1]
41. Are the results of subak meetings usually followed and carried out
    by subak members?
    (a) always [3]
    (b) mostly [2]
    (c) seldom [1]
42. Are there frequent technical problems in the subak, like water
    shortages, pests, and low production?
    (a) seldom [3]
    (b) sometimes [2]
    (c) frequently [1]
43. Are there frequent social problems in the subak, like conflict
    among members, water theft, failure to work together or to care
    about religious rites in the subak?

(a) seldom [3]
(b) sometimes [2]
(c) frequently [1]

44. In your opinion, is there a connection between the strength/capability of the subak and caste conflicts within the subak?
    (a) seldom [3]
    (b) sometimes [2]
    (c) frequently [1]

45. In your opinion, is there a connection between the strength/capability of the subak and differences in the level of prosperity or poverty of subak members? *(Note: we refer to this as "class")*
    (a) seldom [3]
    (b) sometimes [2]
    (c) frequently [1]

46. The capability of the subak members to overcome difficulties, whether technical or social, is:
    (a) very capable [3]
    (b) capable [2]
    (c) not very capable [1]

47. Imagine there are two candidates for subak head. Wayan really wants to be the head. Ketut will only serve if asked by the members. Who is the best choice?
    (a) Wayan (if not selected, the subak could become hot)
    (b) Ketut (because it is easy for the subak to believe in him)
    (these were not converted to a numerical scale)

48. Who is best to become a subak head?
    (a) a rich man
    (b) an average man
    (c) a poor man
    (d) all are OK
    (these were not converted to a numerical scale)

49. Who is best to become a subak head?
    (a) a high-caste man
    (b) an ordinary-caste man
    (c) all are OK
    (these were not converted to a numerical scale)

50. Are subak members embarrassed or fearful to be fined?
    (a) embarrassed
    (b) fearful
    (c) neither
    (these were not converted to a numerical scale)

51. Are all the members of the subak of ordinary caste?
    (a) yes
    (b) no, there are a few high-caste men
    (c) no, there are lots of high-caste men
    (these were not converted to a numerical scale)
52. What is (most) damaging to the subak?
    (a) price of rice
    (b) sawah converted to other uses
    (c) local influences (other economic opportunities)
    (d) functional capability of the subak (subak does not work well)
    (these were not converted to a numerical scale)

CPSIA information can be obtained
at www.ICGtesting.com
Printed in the USA
JSHW020418270819
1217JS00002B/6